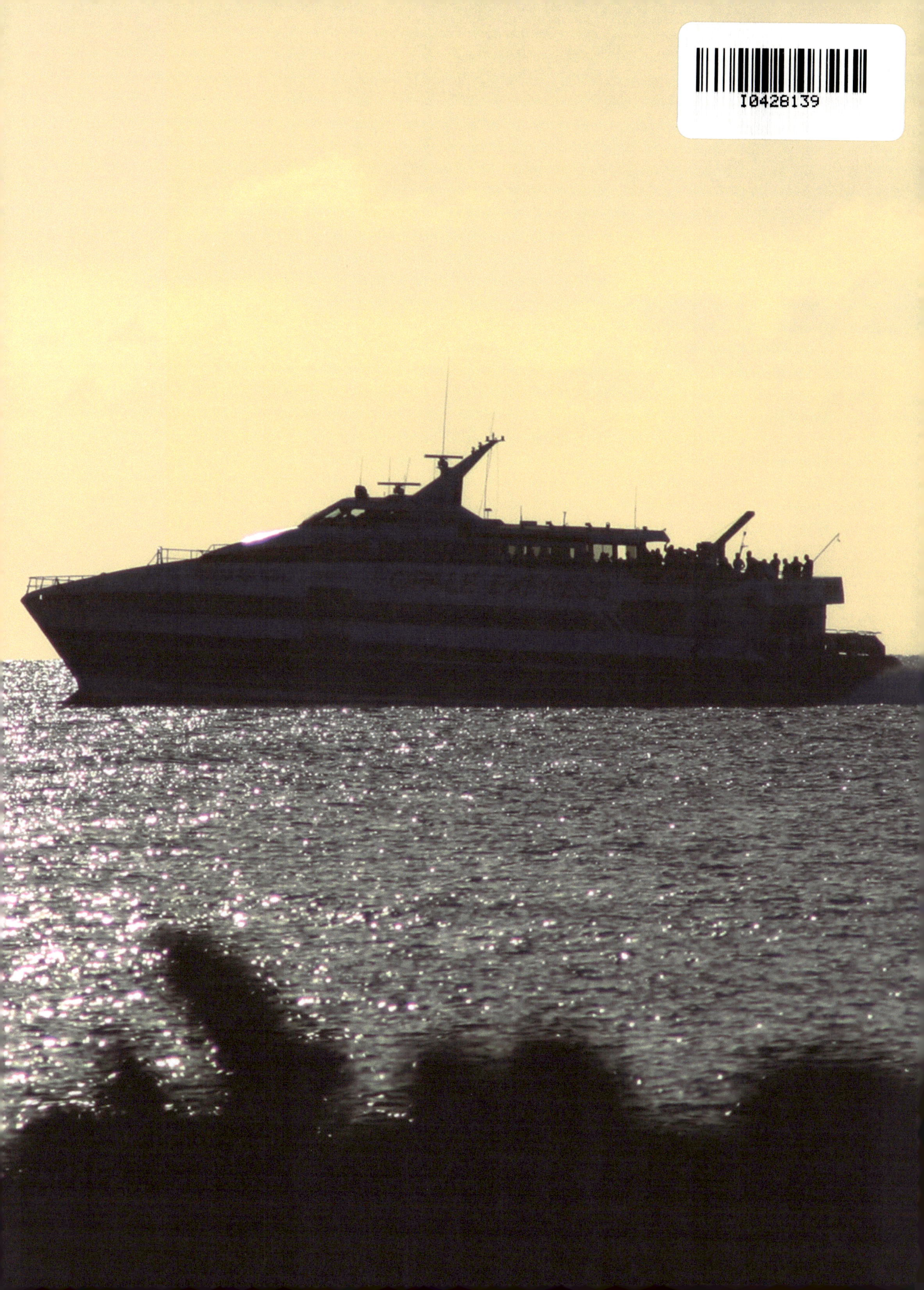

Table of contents

	Introduction	4
1.	The begin - Need for speed	5
2.	Fast across the water	8
3.	The first steps	14
	The Turbinia - how a steam yacht caused a sensation	*14*
	The „Finnjet" - The most exceptional ferry of all time	*20*
4.	Across the water on stilts	23
	First successes	*25*
	The „PT.50" - A new era is beginning	*27*
	The Russian approach	*30*
	America's separate World	*34*
	The Boeing „Jetfoil"	*37*
	What was left	*40*
	When the water police learned to fly	*42*
	Powered only by the wind	*44*
5.	Hovering on air - In the beginning was the hairdryer	49
	England's roaring proud	*52*
	Smaller and finer - The AP 1-88	*54*
	A classic hovercraft	*57*
	New from the old continent - the ABS M-10	*59*
6.	Riding on an air bubble	61
7.	On one, two or three hulls	65
	The Hybrid - a typical HYSUCAT	*76*
8.	The catamaran-boom	78
	Catamaran success in Norway	*81*
	Crewboats - The bus to the drilling rig	*84*
	The biggest crewboats	*85*
	Highspeed boats in use in the wind power industry	*87*
9.	About construction	89
	Like a rocket	*91*
	Magic the power out of the cylinders	*94*
10.	The big fast ferries	99
	INCAT - A Tasmanian success story	*100*
	Progress from Western Australia	*107*
	Japan's Dream - The Super Technoliner	*111*
	First the TGV, then the BGV	*113*
11.	Flying ships - the airborne future?	114
	The „Ekranoplan"	*117*
12.	High-Speed for modelmakers	120
13.	Finally	122

Introduction

This is the fourth edition of my book „The World of Highspeed Ferries", which was first published in 2007. It contains some additions and updates. The topics of the wind-powered „foilers" and the new electrically powered high-speed ferries are presented here for the first time.

The reduction of CO_2 emissions is becoming an increasingly important topic in shipping. Likewise, the IMO has prompted merchant shipping to take extensive measures, such as retrofitting filter systems and new propulsion systems, by imposing considerable limits on the emission of sulphur dioxides and nitrogen oxides. Tonnage that can no longer be retrofitted is increasingly being scrapped - a fate that also threatens obsolete high-speed ships.

New fuels and energy sources are coming to the fore. From small high-speed water taxis powered electrically by batteries to massive catamarans that are to run on hydrogen, everything is possible again. Likewise, the sailboat hurtling through the water on hydrofoils has become a fixture in international sailing. Lightweight construction and new technical discoveries made in recent years have led to a renewal of the world of high-speed vessels.

The history of high-speed ferrying began its rise directly after the Second World War with exotic and at that time absolutely novel vehicles such as hydrofoils and hovercrafts. The idea, very common among technologists at the time, that anything was possible in technology if you just wanted to achieve it, extended to this sector as well. In the ideas of the time, many foresaw the emergence of very large ships speeding along on hydrofoils or on air over water.

Today, the knowledge gained throughout history is not only dedicated to saving travel time, but also contributes concretely to reducing the planet's CO_2 load. One can therefore speak of a new revolution in shipping, which not only encompasses short-distance transport as before, but also spills over into the world of cargo ships.

This book aims to provide its readers with an overview of the background, historical development and technologies of this shipping sector. Selected examples of the various technologies and historical events significant to the industry are presented in this book. In particular, those have been chosen which have prevailed in competition or which stand out from others due to their special features.

C. Schramm - July 2023

1. The Begin - *Need for speed*

The need to race faster and faster seems to be an inner compulsion for human beings. The constant pursuit of world records shows this, as the achievement of a new top speed is usually of no particular benefit to the general public. Only the personal satisfaction of the record holder seems to count.

This book, however, is about top speeds achieved solely for economic benefit. In today's global mass society with its complex network of transport links, the need to save time has become vital.

Once upon a time in the 19th century, steam-powered locomotives were the fastest land vehicles. At sea, first sailing clippers and then the new steamships reached higher speeds than ever before. In Nelson's time, a frigate could sail at 8-9 knots in a good wind, but clippers could reach twice that. Steamships could eventually reach more than 30 knots with the introduction of steam turbines. But it was the aeroplane that made speeding possible.

It was not until the second half of the 20th century that merchant ships were built that could go more than 20 knots in normal operation. Today, the high price of oil and the need to produce fewer exhaust gases are driving ships back into the speed range of the old pre-WW1 cargo steamers. Nine knots is not at all an unusual cruising speed for typical cargo ships, except for the oversized container ships.

However, normal ferries, such as those in the Baltic Sea, travel between ports at more than 20 knots. Great efforts are made to operate this as energy-efficiently as possible.

This book is dedicated to a special type of merchant shipping, the fast ferries. An important condition for them to be mentioned in this „fast book" is that they must be able to reach at least 25 knots service speed. Their cargo must consist of people and/or rolling cargo - i.e. cars and lorries. Fast container freighters or fast

Fast crossing: Two ferries of the Fred Olsen Lineas race past each other somewhere near the Canary Islands. The summed speed is about 130 km/h. Both catamarans are about 96 metres long.

conventional passenger ferries are only marginal examples here.

Fast ferries are a type of ship that has only developed intensively in the last 30 years or so. They achieve performances and use technologies that were still science fiction for engineers of the 1950s. But today, floating structures made of aluminium are welded together that are several times larger than the first steel steamships. The drives achieve enormous performance with fantastically low weights.

An old ferry from the 1960s, which could carry about 120 cars and about 600 people on board at an average speed of 18 knots, had to

had to move a total mass of between 10,000 and 15,000

tonnes through the water. It was made of steel and its technical systems were large and heavy.

In comparison, a barely 90-metre-long catamaran ferry with the same capacity is downright filigree, but it can accelerate to 35 knots because its consistent lightweight construction means that its total mass is only just under 1,400 tonnes. Its transport capacity is many times greater than that of the old ferry, because it can offer many more crossings on the same route in one day. The technology

(Above) A flotilla of American speedboats speeding across the Pacific in calm seas during the Second World War. The plywood hulls could no longer withstand the shock loads even at low wave heights, so it was only possible to go really fast in good weather and calm waters. In fact, they only went at „full speed" during battles, because the petrol engines had an enormous consumption. The boats reached top speeds of up to 43 knots, but quickly wore out.

(Below) After a quick crossing, a strange vehicle pushes itself onto the beach - a hovercraft. The amphibious AP 1-88 is one of the most successful hovercrafts in the world and was produced in Cows on the Isle of Wight. It is used in all sorts of parts of the world. It will be described in detail in a later chapter.

has not yet been exhausted. In fact, its triumphal march has only just begun. So the next revolution in shipping is still in full swing.

It will bring novel fuels such as natural gas or hydrogen, but also the use of stored electrical energy. Perhaps future high-speed ferries will not race quite as fast as those of the past. But they will be much more environmentally friendly.

(Above/Left) The „Stena Voyager", a giant catamaran powered by gas turbines, leaves the Irish Holy Loch at a speed of almost 40 knots. The ship was scrapped in 2013. Until then, she was one of the most powerful civilian ships in the world. She also holds the world record for the largest catamaran.

Length:	*126.6 m Width: 40.0 m*
Draught:	*4.8 m Passengers / cars 1,500 / 375 cars*
Propulsion:	*2 x gas turbine LM 2500 and 2 x LM 2500*
Total power:	*80,905 kW (108,495 hp)*
Speed:	*40 Kts*

(Below) A Russian hydrofoil of the type „Metora" was lifted onto the quay wall before the winter set in. The aluminium hull and the wings would not be able to withstand the winter ice.

2. Fast across the water

(Above) Powerboat races often show the danger that water poses to fast objects. Just one wave can throw the boat off course. When it hits the surface, the relatively dense water has just as devastating an effect as a concrete wall. People thrown out of such boats often suffer serious injuries.

Nowhere is speed so easily felt as on the water. The stresses of pushing through the waves to which a fast motorboat is subjected are transmitted undamped to its crew. If you want to achieve a similar effect on land, you should drive an off-road car over an extremely uneven track. Most land vehicles can do little to counteract the impact loads that occur and are at extreme risk of suffering serious damage or crashing due to loss of control.

Since water, unlike solid ground, is a more compliant element, humans were able to find solutions to reduce the loads on water vehicles very early on.

Small watercraft that can reach speeds of up to 45 knots have been built for more than seventy years. But large commercially viable vessels capable of comparable performance have only been on the rise for about 20 years.

In what is probably a unique exception in the history of technology, it was not war and the upgrading of propulsion engines that drove this development, but the need to create fast and time-saving connections. So far, the superfast ferries have not yet reached the sizes that are considered normal in container shipping and other branches of merchant shipping. But they are well on their way to entering these size categories as well.

To be able to travel fast through the water, a lot of energy has to be applied, because the resistance of the water increases as the square of the speed. Therefore, some resourceful minds have worked in the past to reduce the amount of energy to be used for this purpose and are still

working on it all the time. The second big problem for designers is that there are rarely water surfaces on the sea that are as smooth as on a duck pond.

The waves on the water are the main problem. The technicians counter this problem with various solutions that have made it possible to achieve remarkable speeds. The following comparison shows how successful this development has been:

In the 1950s, the US Navy had the keel laid for the aircraft carrier „USS Enterprise" (CVN 65), which is the most powerfully motorised ship in world history to date: the full propulsion power of all four steam turbines, which were powered by eight nuclear reactors, amounted to about 360,000 hp! The Enterprise was thus so overpowered that it was feared that damage to the hull would occur if the full power was used. The ship never had to use the full power, because it already reached a full 45 knots when using about 80% of the power. This was the fastest speed the heavy hull could have reached anyway. Physics simply did not allow for much more.

The „Big E" was a giant of more than 45,000 tonnes displacement at a length of 346 metres. But it could only achieve this performance in relatively smooth seas. Even with waves of more than four metres, such strong vibrations and shocks occurred that the sensitive equipment on board, such as the radar electronics, were considerably disturbed. For a long time, the USS Enterprise was the fastest large ship on the world's oceans, but this was kept strictly secret. Even today, the performance of the „USS Enterprise" is published in the official data of the US Navy as „30+ knots". In comparison, the fast ferry „Mastercat", built in Tasmania in early 1998, can also reach 45 knots, but is only about 91 metres long. Nevertheless, a high speed can be maintained even in waves up to 4 metres high. This is made possible by the new special hull shape and stabilising fins attached to the bow. The vibrations and ship movements are so low that passengers can feel comfortable on board undisturbed by seasickness.

The sophisticated technology of this catamaran, which is still light despite its mass of about 1,400 tonnes, ultimately proved to be the superior solution compared to

the use of brute force of the giant. In the meantime, the civilian fast ferry industry has far surpassed the defence industry in the development of fast watercraft in terms of technology, performance and size. A comparative look at the now technically obsolete and largely decommissioned fast boats of the German Navy and catamarans of the same size for passenger transport clearly shows this.

While civilian fast ferries can still operate at full speed in bad weather, the military speedboats already have to slow down in order not to suffer any damage. It took intensive research and development for these performances to be achieved.

Although unusual-looking fast boats were developed and tested before and during the Second World War, their use in civilian industry did not begin until after the last war. This book reports almost exclusively on the peaceful use of high speed. More than in any other branch of shipping, the development of high-speed ferrying is directly dependent on the advances achieved in hydrodynamics and propulsion systems. Thus, the appearance of a new innovative hull form usually also indicates a change in

trend. First came the hydrofoils, then the monohulls and the catamarans. Now the change to the trimaran, the ship with three hulls, is emerging. The hovercrafts emerged as a temporary competitor.

On average, smaller fast ferries today reach service speeds of 27 to 33 knots, while the large RoRo fast ferries can still travel at an average speed of 35 knots. The limit of 40 or even 45 knots is only reached or exceeded by exotics and under special conditions.

It is very likely that the performance limits of water-displacing vehicles have already been reached today. But new solutions are also emerging that blur the dividing line between aircraft and ship. Novel ground effect aircraft, as they are called, have the potential to raise the achievable speed over water to over 200 knots.

(Below) In the 1960s, the USS Enterprise was the pride of the US Navy as the largest warship in the world. However, only a few insiders knew how fast it could really travel. The ship was still in use until 2017. It was then also the longest-serving active warship of modern times. It will most likely be scrapped. The disposal of the eight reactors is a major problem.

(Above) The 91-metre-long „Master Cat" enters a Norwegian port. A rough sea in the Skagarak cannot harm her on the crossing to Denmark. Here you can see the special bow shape known as the „wave piercer", which makes the ship highly seaworthy. The „Master Cat", formerly called „Cat-Link V", is still the record holder in the battle for the Hales Trophy for the fastest Atlantic crossing. Her current name is „Skane Jet".

(Below) A new large catamaran for the US Navy, where it will be used for the rapid transport of material and troops. At 101 m long, the new „Spearhead" class ship is no dwarf. (Width 28.5 m, draught 3.83 m. Speed: max 43 kn.)

(Above) Of course, fast ferries can also jump. But it is not advisable to stand on deck at such a moment!

Which is the fastest ship ? That is certainly an interesting question for many people. However, the operators of these vessels have a clearly defined and completely different requirement: how does high speed on the water help them to make a lucrative business?

The answer to this is mostly a compromise between the capacity, the cruising speed and the necessary energy input for a vessel - not to mention the capital input. Not every solution that appears elegant on the drawing board is suitable for shipping operations.

For example, many enthusiasts still wonder why the genre of hydrofoils has almost died out or why the operation of the famous hovercraft canal ferries was discontinued in the 2000s. For the experts, however, the case is clear. The operating costs were far too high.

The fast ferry business is risky. Often only companies that draw their main income from other sources can afford to enter this industry with due caution. If it does not work out, the resulting loss can still be used to save tax. However, companies that tie their entire economic fate to fast ferry shipping are subject to a high mortality rate. Only about every fourth or fifth company of this kind has survived in the last decades of the

ferry history has survived its first years of life. What breaks the necks of the beginners?

Excessive expectations of passenger potential and underestimation of the necessary effort are at the top of the losing list. Even if, in the eyes of a group of investors,

a new fast ferry line appears to be the best solution to a particular traffic problem, this opinion is not necessarily shared by the broad majority. But calamities such as wage disputes, legal disputes or sudden increases in fuel prices can also put an end to a project.

In order to be able to make qualified statements about the need for infrastructure in a region, years of broad-based studies are often necessary. Private companies often spare themselves this and wonder why so few passengers show up at the landing stages.

Fast ferries are much more prone to technical problems than other commercial vessels. The need to be able to reach high speeds in an economically viable way pushes the technology to the limits of what is justifiable.

Thus, novices in the industry very often underestimate the effort that has to be invested in maintenance on any given day. Other stumbling blocks arise from environmental protection, questions of traffic law and also in problems of having the borderline vehicles of the fast ferry operators approved by the authorities responsible for safety issues - a fate that mostly befalls those projects that are planned on the basis of older second-hand vehicles where modern safety standards can no longer be achieved.

Whatever the concept and the plan, in all cases a lot of money is needed at the beginning to survive the first years of operation. There are a lot of large-scale projects in this industry that have quickly suffered shipwreck - and that is not meant proverbially. The causes are usually to be found in details that were not considered by the planners beforehand. For example, even delays in the granting of operating licences, which only take several weeks, can trigger a kind of catastrophic cascade effect on the rest of the project. If a newly built, expensive ferry has to sit idle in port for several weeks, it must still be possible for the operator to pay the charter and maintenance costs.

A disaster of a different kind for many ferry operators was the construction of the Channel Tunnel between France and England. This long loss-making company has only been in the black since 2017. However, it was able to attract much of the transport volume on this route.

However, the cost of a crossing for the end consumer did not fall in the process. Therefore, a few years ago, a venture capital financed company that put a second-hand fast ferry into service was able to hold its own for some time in the English Channel by offering low prices.

Another success story is this: In the nineties, some enthusiasts interested in hydrofoils started a small fast ferry line in the vicinity of Amsterdam, which was initially operated with an obsolete Russian hydrofoil of the type „Meteora". At first, sightseeing tours were offered on the Ijsselmeer near Amsterdam, but then they applied for a subsidised service on the North Sea Canal between the port of Amsterdam and Velsen-Zuid at the other end of the North Sea Canal.

Holland has an effective system for subsidising local transport links. Instead of handing out money at political discretion, each transport company has to openly declare its passenger numbers and as a result receives a „per capita premium" for the transport performance of the previous year.

Since the line of the „Flying Fast Ferries" (FFF), as they called the young company, already operated successfully from the beginning, the state was also quite generous. Later, the local transport company Connexxion, which operates on a large scale in the Netherlands, took a share in the line.

Connexxion took a share in the line. Now they bought two old „Voskhod" hydrofoils, which with a single propulsion engine and their capacity of 80 passengers represented an ideal scale for the line. The two boats now ran very reliably up and down the North Sea Canal and carried many passengers every day.

Later, Flying Fast Ferries surprised the trade by purchasing three new Voskhods to replace the two older boats. For many years, they operated on the route with their mix of proven Russian technology and modern Western equipment. Business turned out to be to the satisfaction of the investors, and the traffic-congested surroundings of Amsterdam were relieved a little.

The secret of FFF's success lay in its cautious approach in the early days with ships that could be procured cheap-

(Above) A „Voskhod" hydrofoil of „Flying Fast Ferries" from the Netherlands. The boat is very economical and robust in operation.

ly. The Russian hydrofoils, once one learns how to handle them, are reliable and extremely economical vessels.

Nevertheless - as is so often the case - a public rethink brought an end to this. In 2014, the permitted speed on the North Sea Canal near Amsterdam was drastically reduced. Operation no longer made sense.

View from the front passenger compartment of a „Voskhod". The view is very tourist-friendly.

3. The first steps

The Turbinia - how a steam yacht caused a sensation

(Above) Parsson's steam yacht „Turbinia" surges ahead. It is hard to imagine how hard the poor stoker had to slave. He had to fill three fire hatches with coal in the narrow boiler room alone, as best he could. The yacht's cruising distance was therefore limited less by the coal supply than by the condition of the engine crew.

In the middle of the 19th century, the steam engine had almost completed its triumphal march in the shipbuilding industry. More and more cargo and passenger steamships were built, but they still travelled quite slowly. Cruising speeds of around 9 knots were typical for the time.

This is because the energy utilisation of the steam pressure by a piston steam engine was still rather ineffective. Although it was possible to improve the degree of performance with the triple and later the quadruple expansion engine, with three and four cylinders respectively, there was still a narrow limit to the highest achievable speed of the engines. Anyone who exceeded this risked serious damage to the complex mechanics of the drives. To control the piston movement, a complicated linkage had to move the valves in sync with the piston movement.

Progressively, it was possible to build more efficient steam boilers that could cope with ever higher steam pressures. The time had come to invent a new prime mover to harness this power.

Although there were several researchers working on

Although there were several researchers working on turbine wheels as a replacement for the steam piston, the honour of having built the first practical steam turbine into a ship fell to the Englishman Sir Charles Parssons. He financed the construction of an experimental yacht,

the small „Turbinia", out of his own pocket.

The idea of the turbine wheel was inspired by windmills and the water turbines already in use at the time to generate energy. Inventors like Sir Charles were convinced that this effect would also work with the steam jet of a boiler. Parsson's decisive contribution was that several turbine wheels, increasing in diameter, were mounted on a shaft. This meant that cooling and expanding steam could be used economically. The Parsson turbine has become the most widely used turbine design. Its principles live on in today's jet engines.

The English Admiralty was initially not very receptive to this new technology, so Parssons decided on an offensive „marketing strategy":

He smuggled himself into the ranks of spectator yachts with his small ship „Turbinia" at the fleet parade at Spithead on 26 June 1897, broke away from this group at a suitable time and sped past the royal yacht and the parading fleet at a speed of almost 34 knots. The excitedly swarming guard boats were unable to catch him.

Whether this was an unannounced demonstration or a concerted manoeuvre has never been known. But instead of a conviction for gross mischief, Sir Charles Parssons achieved the desired success. His invention, the steam turbine, was installed in two new experimental destroyers.

From then on, the steam turbine spread unhindered. It was used everywhere where a high cruising speed was required, such as for example

the fast Atlantic steamers. The „SS Mauretania" was the first of these giant ships to achieve the „Blue Riband" at a speed of about 26 knots with the new technology. Before the introduction of the Parsson turbine, a ship with 20 knots was already considered very fast, but after the introduction of the new propulsion, the ships reached more than 35 knots in time. In those days, however, the world's international navies were ahead in terms of speed.

The speed of the old fast ships depended primarily on two factors, namely the slenderness of the hull and the en-

gine power. The more powerful the engine plant became, the heavier it also became, and in turn the larger the hull had to become in order to still allow for a sufficient degree of slenderness.

The length of the fast passenger steamers was soon almost 300 metres. These giants often had more than 120,000 hp of propulsion power and burned incredible amounts of coal and later also fuel oil in their boilers.

For an Atlantic crossing from Southampton to New York, several thousand tons of coal were consumed in about four to five days of sailing. On the SS Mauretania, several hundred stokers toiled with shovels to feed the boilers with about 1,000 tons of coal every day. The zeal to save travel time at any cost had taken hold of humanity. When crossing the Atlantic, it became extremely impor-

tant for merchants and tourists to save even a single day of travel on the routes. Routes that had been sailed by the old wooden merchant sailors in only many weeks or even months of travel time. The speed revolution had begun. It did not only affect the maritime world at that time, for faster and faster locomotives and land vehicles were also being tried out on land.

But in those days, the cost of fuels like coal and the labour costs of stokers were so incomparably low that the effort was worthwhile in transport between the economic centres of the USA and Europe. Nevertheless, there were technologists who clearly recognised the inefficiency of the large steamships, as they required an enormous input of energy to fully develop their performance. Thus, around 1877, the engineer John Isaac Thornycroft began

The „SS Mauretania" under steam. She was the first large passenger ship in the world to be powered by steam turbines.
She thus established a new phase in the race across the Atlantic, as she was once again able to shorten the journey time by several hours. I
her sister ship, the „SS Lusitania", unfortunately became best known for her sinking by a German submarine on 7 May 1915, which served as a political reason for the USA's entry into the First World War.

Length:	240.5 m
Width:	27 m beam
Tonnage:	31,938 BRT
Propulsion:	4 steam turbines
	25 boilers
Power:	78,000 hp (58 MW)
Speed:	26,5 kn
Passengers:	563 (1st class)
	464 (2nd class)
	1,138 (3rd class)
Year of construction:	1906

The „SS Mauretania" passes a lighthouse on the Channel coast during her sea trials in 1907.

to consider how to reduce the resistance of the water.

Even slimmer hulls would inevitably have led to less stability - with capsizes as a consequence. Thornycroft came up with the idea of hollowing out the hull at the bottom of the ship and filling it with air. The ship would thus float on an air bubble, which developed less frictional resistance than the water on the hull. In model tests he proved the validity of his claims. But what worked excellently in the model with spring drives, bellows and miniature steam engines on an experimental pond could not be implemented on a large scale at the turn of the 20th century. The power of the available drives and blowers was simply not sufficient. Nevertheless, this was the birth of hovercrafts Very many decades later .the Vosper-Thorneycroft company actually built a rather large hovercraft.

Many decades later in the 1960s, Vosper-Thorneycroft built the large VT-2 hovercraft in England as a prototype.

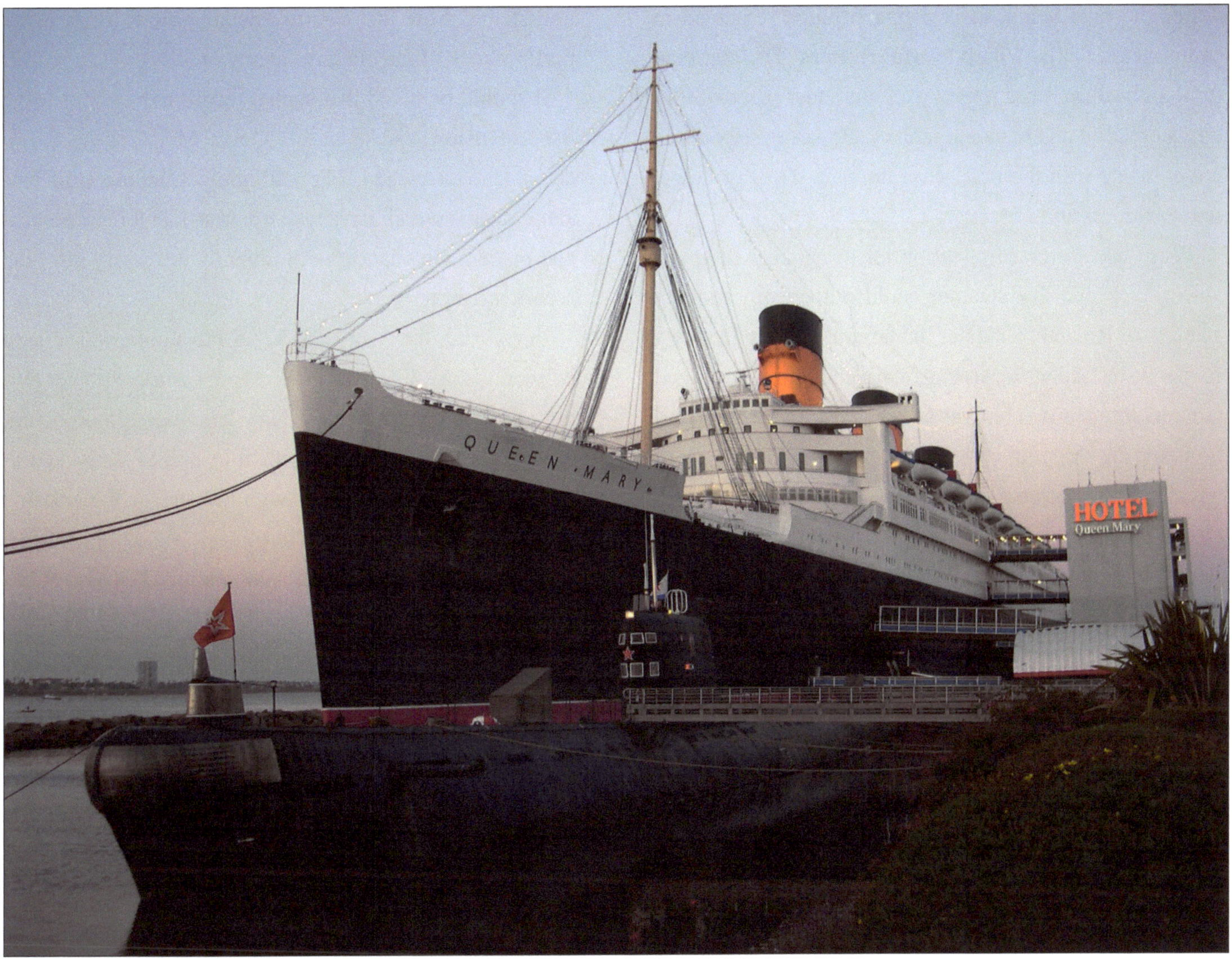

(Above/below) The „SS Queen Mary" is now moored as a hotel ship at a pier in Long Beach. She is the last surviving evidence of the Transatlantic Race. Commissioned in 1936, she was the pride of Great Britain. She survived the Second World War unscathed as a troopship. Today she is the last remaining example of her era.

After her record-breaking voyage across the Atlantic in 1936 at almost 33 knots, she set a record during the war for the largest number of passengers on board a ship: during a crossing in 1944, she carried more than 11,000 soldiers.

Length:	310.7 m
Width:	35.46 m
Measurement:	81,235 BRT
Propulsion:	4 geared turbines
	24 oil-fired boilers
Power:	approx. 200,000 hp
	(149 MW)
Passengers:	2,280 in three classes
Max. Speed:	33 - 35 knots

By Mfield, Matthew Field, 2008

For the time being, the existing solution remained to equip slender hulls with powerful engines. The marines' torpedo boats and destroyers were the most successful in terms of speed and manoeuvrability. However, they could not compete with the first ocean liners in terms of range, seaworthiness and capacity.

There was fierce competition for every knot of speed with which one fast steamer could outperform another. The „Blue Riband" and later the famous „Hales Trophy", for the fastest Atlantic crossing of merchant ships, changed winners frequently. This was associated with great reputations for the shipping company and also for the nation to which the ship belonged.

The races between steamships on the Atlantic were reported in detail by the media. They were thus an important object of public interest. All in all, this competition ensured a rapid and far-reaching introduction of the various technical advances in the field of ship propulsion, hull construction and ship safety technology.

It is certainly true that the construction of such super battleships as the „Bismarck" and the „Yamato" and the construction of the powerful aircraft carriers of the US Navy in the Second World War would not have been possible without the experience gained from the Transatlantic Races.

While the history of this competition is not the subject of the book, it is interesting to see the development of size and performance in naval engineering that resulted from the competition. Next to a giant ship like the old „Queen Mary", today's catamaran ferries like the „Hoverspeed Great Britain" and the „Cat Link V" look like tiny dwarfs. But thanks to the construction of compact diesel engines, the lightweight aluminium construction and the advanced

hull shape, these fast catamarans are able to surpass the performance of the old days at any time.

It should be noted that such a record trip did not leave the propulsion systems of the „Queen Mary" or the „United States" unscathed. The extreme continuous performance over several days caused considerable problems and made maintenance work urgently necessary after the record attempts.

In contrast, the „Cat-Link V", for example, was able to deliver its cruising performance with a routine power output of 90 percent. Since the ferry had no cargo on board, its draught was slightly less than in normal ferry operation. This difference was enough to increase the cruising speed from about 38 to 41 knots.

There will probably never again be such fast luxury liners as the „Queen Mary" or the „United States" - the energy expenditure and the technical problems are too high to be justified in any way economically or environmentally.

The future belongs to the smaller and lighter racers. Even in container shipping, it is unlikely that large ship units will travel faster than 25 to 28 knots in intercontinental traffic. Today's passenger giants, which are still being built much larger as cruise ships

than the old transatlantic steamers, rarely reach speeds higher than 23 knots.

For a few years, a kind of „arms race" prevailed between large shipping companies in the Greek island world and on the long ferry routes across the Mediterranean. These giants drove up the speed of conventional ferries, which had been the norm until then, from about 21 to almost 30 knots.

The price for this was extreme fuel consumption, which was of course passed on to the ticket prices.

Ship	Year	Power HP	Speed	GRT
SS Sirus	1838	320	8.03 Kts	703 GRT
SS Kaiser Wilhelm II	1904	44,500	23.58 Kts	19,361 GRT
SS Queen Mary	1936	200,000	30.63 Kts	80,744 GRT
SS United States	1952	240,000	35.59 Kts	53,369 GRT
Hoverspeed Great Brittain	1989	19,579	36.97 Kts	ca, 1,200 GRT
Cat Link V	1998	37,977	41.28 Kts	ca, 1,900 GRT

(Above) Left alone - the 301-metre-long „SS United States", built in 1952, was a celebrated flagship of the American merchant navy in the 1950s.
Today, she lies in a shabby condition in the port of Philadelphia, awaiting her still uncertain fate. When she was built, she was lined with asbestos in every conceivable place to secure her against fires. This material, which is now strictly forbidden in shipbuilding, is a heavy legacy that can only be disposed of at great expense.

(Above) „Hoverspeed Great Britain", an Incat-74 metre class catamaran, leaves New York harbour in June 1990 to win the Hales Trophy for the fastest Atlantic crossing with an average speed of 36.97 knots after three days, seven hours and 54 minutes. The trophy, which is administered by the New York Maritime Museum, had to be handed over by court order after the record-breaking voyage. Since then, only catamaran fast ferries have been able to surpass this performance. The current record holder is the „Cat Link V" with 41.28 knots, which is now called „Master Cat". It is not expected that a conventional ship will ever break these records. Only a special construction like a lightly built catamaran is capable of doing so.

(Right - top picture) The SS „Zealand-Market" in 1981. The SL-07 container ships were the fastest container ships of all time. With a power of 120,000 hp from two steam turbines, they reached a speed of 33 knots with an extreme fuel consumption of 614 tons per day. Built in the early 1970s, these eight ships were sold in 1981/82 to the US Navy, which gleefully welcomed them as a new class of heavy high-performance transport ships. They were converted into RoRo carriers to carry the equipment of complete armoured divisions of the US Army.

(Right - lower picture) The former „Sealand Commerce" after her conversion to a fast RO/RO transport ship. She is in the service of see Military Sealift Command of the US Navy. The seven units of the Algol class can each carry about 700 military vehicles. Their total payload can already tip the balance in a hypothetical serious conflict between the USA and China or Russia. In the process, they would race across the Pacific or the Atlantic in just a few days. Their strategic importance is very substantial.

The „Finnjet" - The most exceptional ferry of all time

The Finnish shipbuilding industry has built many an unusual ship over the last 50 years, but with the „Finnjet", which entered service in 1977, it created a technological monument of a special kind. At a time when an ordinary ocean-going car ferry could only manage 20 knots and was usually not particularly large, the Finnish shipyard OY Wärtsila in Helsinki came up trumps with a miracle ship that far outshone anything that had gone before.

The ship, which was built on behalf of Finnlines and was over 212 metres long, was not only larger than most other ferries, but its propulsion system, hull shape and the design of its superstructure were also completely new.

The „Finnjet", which travelled at about 30 knots in normal service, was powered by two Pratt & Whitney gas turbines, each with an output of 27,500 kW or 36,878 hp, acting on two controllable pitch propellers. The hull corresponded to the highest Finnish ice class, because the giant ship was supposed to be able to cross the often frozen Baltic Sea in winter without any significant stop.

The elongated, black-painted hull adjoined bulky superstructures that were bordered almost only by right angles. Instead of playfully emulating the elegant

Instead of playfully emulating elegant yacht building, the „Finnjet" had a modern, timeless, industrial design that radiated power and unstoppability. To this day, the „Finnjet" would still be able to steal the show from other cruise ships or ferries.

The turbine exhaust gases were emitted through two chunky yet sleek funnels that had been custom-fitted to the ship. Even the sound of the ship's engines somehow sounded different from other ships: Instead of a dull whump or thump, the muffled whine of jet engines could be heard behind the funnels on deck, swelling on and off depending on how much power was being demanded at any given moment.

The new ship could carry up to 1,532 passengers and 350 cars, as well as a large number of trucks. This far exceeded the capacity of other ferry vessels. With the „Finnjet", Finnlines replaced an older ferry that had still required almost 1 ½ days for a journey between Travemünde and Helsinki. The Finnjet reduced this journey time to 22 hours. In order to be able to use the high-performance ship optimally, everything had been coordinated for a short turnaround time in the port. The Finnjet was to be able to cast off again just 120 minutes after docking and set off in the opposite direction. Wide ramps, a sophisticated ramp system and optimally planned logistics made this possible.

The supplies for the galleys and the goods were delivered in containers and were ready to be taken on board when the ship arrived. A small gantry crane housed in the superstructure behind the funnels now exchanged the containers without having to disrupt the ferry's ramps.

In open water, the „Finnjet" could reach up to 32 knots. But this required a water depth of more than 100 metres, as the suction from the screws was too great in shallow water. At shallower water depths, a suction effect was created between the bottom of the ship and the bottom, similar to that of a jet of air being sucked through a narrowing pipe.

At that time, there were no diesel engines that could have delivered the required power at an acceptable weight. So they entered the then unknown field of gas turbines in merchant shipping and used two Pratt & Whittney turbines, which were basically jet engines converted for marine use for the DC-8 jet.

Each engine weighed only 3,200 kilograms, which was only a fraction of the mass of a similarly powerful diesel engine. A third engine was taken on board as a reserve unit. It was lashed in the engine room between the two active turbines. In the event of damage to one of the operational turbines, a long-term stay in port was no longer necessary, as the engine room crew simply replaced the defective engine with the reserve turbine within a few hours. This could even be done at sea, as only one turbine could already propel the ship forward at more than 22 knots.

During the journey through the icy areas of the Baltic Sea in winter, the ferry could do without icebreaker assistance in most cases. The engines were set to a power-reducing „ice mode", which was sufficient for a speed of 18 knots. The ship sailed the route between Helsinki and

Travemünde for 30 years without an accident. Heavy seas, super-cold ice winters or stubborn pilots could not stop her. Only the gradual increase in the price of oil over the course of her time in service made it difficult to operate her profitably. But the ship's owners, who knew her true value well, did not replace her, but found an innovative solution to save fuel without sacrificing performance.

In 1981, electric traction motors were fitted to the reduction gears of the gas turbines in an Amsterdam shipyard. They were powered by a total of 11,558 kW or 15,500 hp diesel generators. Thus, the „Finnjet" could now either sail at only 18 knots with the new auxiliary drive or reach full speed through the turbines. This saved a lot of fuel during the off-season in winter. The fuel-intensive gas turbine propulsion was only used during the time when most holidaymakers travelled to Finland.

But unfortunately, the old gearboxes did not allow electric motors and gas turbines to be operated on one shaft at the same time. Therefore, until now there were only the two speed levels mentioned. In order to enable flexible mixed operation, the two gear units were exchanged for new equipment in 1994 in an elaborate rebuild.

Since then, the „Finnjet" could go full speed with all available power - the gas turbines and the diesels at the same time - while in winter operation only the diesels were used. The Finnjet underwent several major interior renovations, as the ship's shops, furniture and decorations had to be adapted to the tastes of the times. The ship was painted white in the 90s and switched to the Silja Line.

With the advent of the Greek „Superfast" ferries in the Baltic Sea, the „Finjet" faced strong competition for the first time. As a consequence, it was withdrawn from the Travemünde route in winter and initially operated a winter route to Rostock. But later, at the beginning of the 2000s, the Travemünde business was completely abandoned. The „Finnjet" was therefore put up for sale for about 18 million euros. In 2007, it was used as emergency accommodation for hurricane victims in New Orleans. A new owner could not be found. She was scrapped in India from September 2008. A very sad moment.

The „Finnjet" in facts and figures as last time:

Length:	214.96 m
Width:	25.40 m
Draught:	6.89 m
Deadweight tonnage:	2,728 tdw
Propulsion:	2 gas turbines PW FT 4C-1DLF
Power:	2x 27,500 kW
Secondary electric motors:	2x 7,200 kW
Speeds:	
Diesel and gas turbines:	32.5 Kn
Gas turbines only:	31.1 Kn
Diesel / E-motors only:	18 Kn
Passengers:	1,651
Cabins:	533
Crew: approx.	178
Car load: approx.	325
Truck/trailer load: approx.	50

(Below) Ice travel was one of her great strengths. No other ship as fast as the Finnjet was ever able to do so. The Finnjet was one of the greatest technical achievements of Finnish shipbuilding.

(Above and below) Powerful and elegant at the same time. When the „Finnjet" appeared on the scene in 1977, the design of the hull and super-structure was unprecedented. But even today, her design can hold its own against the gigantic floating hotels of cruising, because she was still a „real" ship - one with very special characteristics to boot.

4. Across the water on stilts

(Above and below) All that remains of Enrico Forlani's boat are old photos and the sketch below. But they illustrate the way it works. You can see the ladder arrangement of the wings and the two propulsion air propellers.

The lesson that emerged from the experience with the first fast ships was clear: if you wanted to save energy, you had to reduce the resistance of the water as much as possible. A slender hull only makes sense up to a certain point: because a narrow piece of ship may be able to float, but it is no longer very stable. In general, there is little that can be done with a hull that is ideal in terms of flow (for example, one metre wide but 40 metres long). Only if you put two of these narrow hulls next to each other and connect them with a platform might it work, This solution is called „catamaran" and will come later.

At first, the solution was: why not leave the water completely? And so various inventors devised methods that could lift a boat hull partly or completely out of the water. There are various methods, but the first successfully applied one was to glide hydrofoils through the water, whose buoyancy lifted the hull.

Among the first pioneers to have practical success with this method was the Italian Enrico Forlani in 1910. In the 1920s, the German Baron Hans von Schertel joined them. Both recognised that the lift-generating wing profiles of aircraft wings introduced in aviation at that time could also be used in water. They would only have to be made smaller. This is because water, which is considerably denser than air, allows for a much stronger lift than in an aeroplane.

Since Forlani was able to have quite powerful petrol engines at his disposal after the turn of the century, he had

suitable sources of power for his experiments. This was a decisive advantage, because even a small steam turbine would have been too heavy. Forlani's first vehicle was a small boat that looked as if someone had attached several ladders to the front and back.

In fact, the ladder rungs were small hydrofoils that were in no way intended for climbing. There was a clever idea behind the rung-like arrangement of the wings:

The boat was calculated in such a way that in its normal state, only a few of the lowest wings were sufficient to lift the hull out of the water at high speed. Should a weight shift occur or a larger wave come along, another rung wing would inevitably be submerged and add its buoyancy to that already present.

Conversely, the buoyancy would be reduced if a wing that was already submerged emerged from the water. This way, the boat would always be stable and would not capsize as soon as the hull lifted out of the water. The principle worked and the hydrofoil was invented.

However, Forlani's arrangement of the wings created too much drag and was too delicate for practical operation to be truly efficient. These first experimental boats, and also their replicas in the USA, were propelled by propellers. This way, the designers did not have the additional

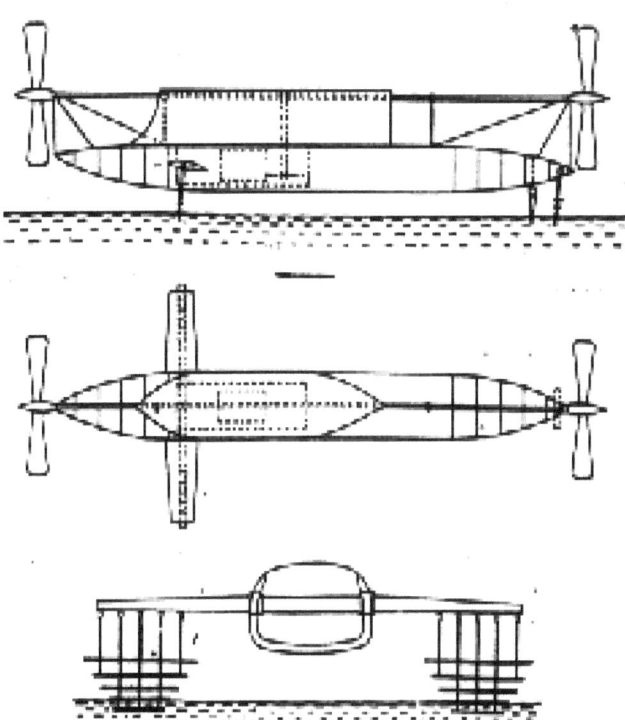

problem of having to develop efficient high-speed propellers.

However, the trend of using propellers for propulsion did not continue. Only in the 90s did a new attempt with this propulsion system appear in San Francisco. The extreme noise and the further dangers of this drive brought the attempt to a quick end.

(Above) Alexander Graham Bell, after giving the world the telephone, read about Forlani's experiments and built his own hydrofoil in the USA. With it, he achieved a first world speed record for watercraft.

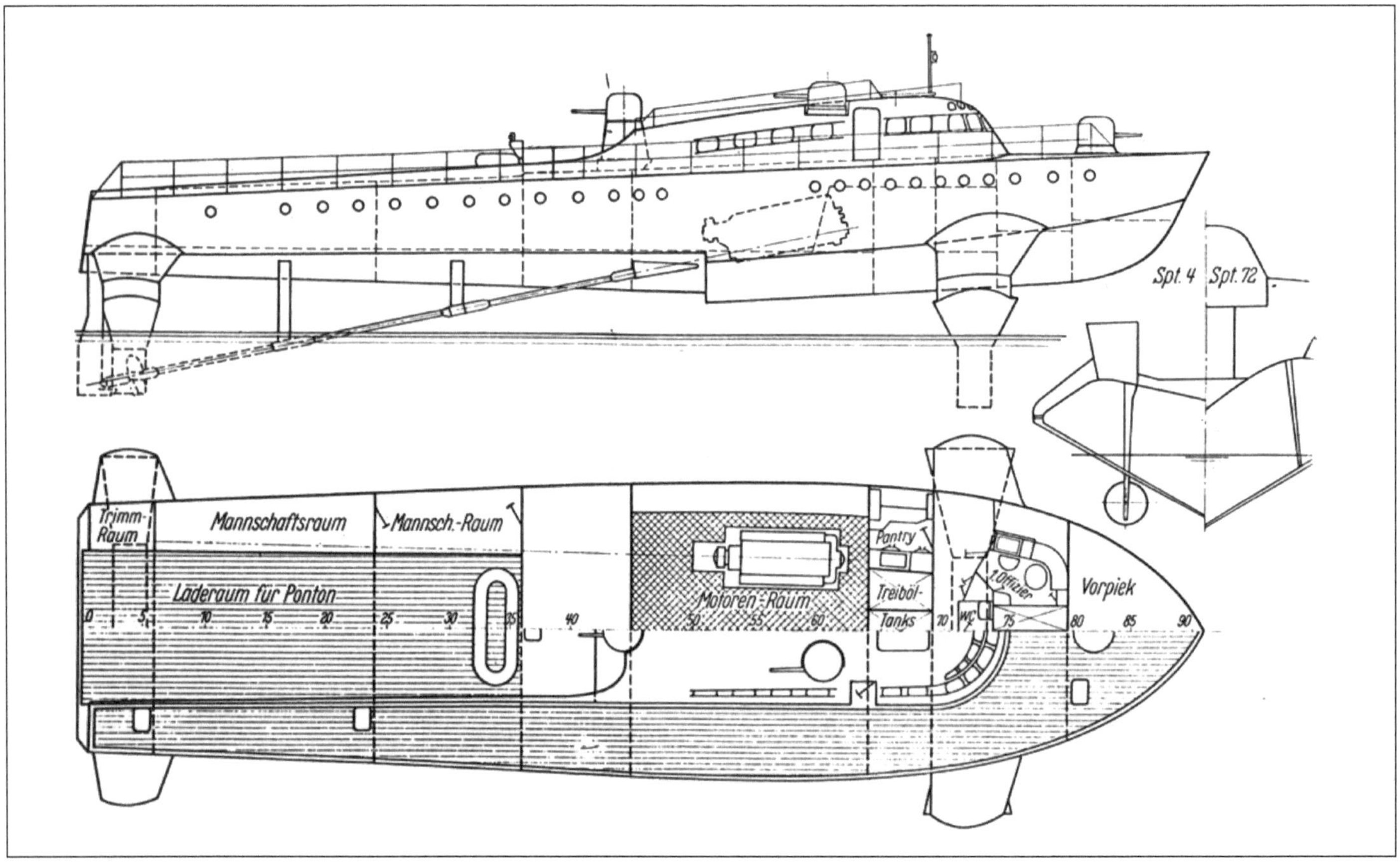

(Abov) „Schell 1" or also „VS 8" was a military transport hydrofoil built in Germany and far ahead of its time in terms of performance and size. Its history is described on the following page.

Data:

Length:	*31.9 m*
Overall width:	*10.26 m*
Propulsion:	*2 x diesel engines MB 501 (1,491 kW each)*
Speed: approx.	*37 Kts*
Year of construction:	*1943*

In the 1930s, the engineer Baron Hans von Schertel, together with his colleague Sachsenberg, researched another design: the V-shaped wing arrangement. Thus, their wings formed an approximate V-shape. Part of the wings penetrated the water surface.

This had several benefits for the stabilisation of the boat: The boat was stabilised in its „flying height", because when a higher load occurred over a wing, for example due to an increased load, the wing dipped a little deeper into the water. Consequently, part of the wing surface that had previously still been sticking out of the water was also used to generate lift.

If the boat heeled to one side, part of the wing surface protruding from the water would also dip on one side and generate more lift. The boat would right itself again.

This design worked very well. There were privately financed experimental boats at first before World War 2, and later the Kriegsmarine promoted the construction of experimental carriers. The largest of all, the „VS-08", became the first hydrofoil used as a commercial vehicle. Among other things, it carried supplies weighing up to 20 tonnes from Europe to the German troops in Africa until it ran aground on the African coast.

For the first time, a successful design had been developed for a fast ship that did not rely on water displacement to generate buoyancy. The first boom in fast ferry shipping had begun. Few know that others in Schleswig-Holstein were also experimenting with hydrofoils on the Schlei.

The German professor Tietjens built several small experimental boats there in cooperation with the Vertens

(Above) Baron von Schertel built the VS-8 (designated as „Schell 1") for the Kriegsmarine. After extensive trials, it was used as a supply vehicle for the Afrikakorps, which was in distress in 1943.

yacht yard, some of which reached a remarkable 55 knots. Technical and financial problems led to the discontinuation of the trials at the end of the 1930s. Today, unfortunately, there are no longer any traces of this work.

Baron von Schertel could not find sufficient resources to continue his work in post-war Germany and emigrated to Switzerland. There, together with other colleagues, he founded the company „Supramar AG".

In addition to some smaller boats such as the PT.4 for authorities, the first series product as a ferry was the „PT 10". This civilian commercial vehicle for up to 28 passengers now transported passengers back and forth on the alpine lake Lago Maggiore, between Locarno and Arona, for a long time. The „PT 10" reached 38 knots and shortened the previous travel time of the old-fashioned steamers from almost three hours to 48 minutes.

This achievement did not go unnoticed and many experts and journalists from all over the world visited the new line. One of them was the Sicilian shipyard owner Carlos Rodriquez from Messina. He immediately recognised the potential of the technology and saw it as a good opportunity to improve the transport connections of his home island to the mainland. The „Supramar AG" and the „Cantiere Navale Rodriquez" formed an alliance to produce the PT.10 and even larger hydrofoils. The next boat was the „PT.20", which could carry up to 57 passengers at 34 knots with a mass of 27 tons. It was seaworthy and was immediately used in 1956 on the route from Messina to Reggio di Calabria on the Italian mainland.

The „PT.20" was so successful that it was sold in large

(Left) A small experimental boat designed by the German engineer Friedrich Wendell, now on display in front of the German Maritime Museum in Bremerhaven.

It featured one of the first applications of fully submerged hydrofoils, but also revealed the main weakness of the concept, as it did not have an automatic stabilisation system, but had to be steered through the water by the skipper himself, like an aircraft, using a control horn in all three axes. That was quite a nerve-racking activity. Flying low just has its special charms ...

numbers worldwide until 1960, when the even larger „PT.50" for almost 100 passengers appeared. It became the most important hydrofoil of this early period. This type of boat was used as a fast passenger boat from Hong Kong via Australia to Europe - and there from Norway to Greece. Its importance for the development of this shipping line is so great that it is worth taking a detailed look at this boat.

(Below) „Cross-overs" between technical fields occur again and again. Here, a launching aid for seaplanes was tested, but did not prove successful.

Similarly, there is at least one amphibious automobile that can reach a respectable speed on hydrofoils.

(Right and left below) The PT.20 was the first successful series of the Supramar AG. Among other things, it appeared as an escape vehicle for a film villain in the James Bond film „Thunderball".

The „PT.50" - A new era is beginning

The PT.50 hydrofoil became one of the most widely used types of this type in the West. The size and capacity chosen by Schertel was ideal for the needs of the first fast ferry operators. Unlike its predecessors, the craft was sufficiently seaworthy to be operated in coastal regions without significant limitations. Its design was simple and robust.

Made of aluminium riveted together like an aircraft, it chased through the water on two steel wings. The front one provided about 80 % of the required lift and was shaped in the typical V-shape of Supramar. This stabilised the vehicle against heeling in the manner already described. The smaller rear wing was only slightly V-shaped and also served as the rear attachment for the two long propeller shafts. These led from there at an angle of about 15° for-

ward to the bottom of the ship. About in the middle of the hull they entered the engine room and were directly connected to the two MTU-12 cylinder engines. Complicated propeller systems had been dispensed with and only two non-adjustable screws were used, which rotated at a very high speed. In order to be able to drive backwards, the direction of rotation of the propellers was changed with the reversing gears attached to the diesels.

The MTU engines, each with a power output of about 950 kW, were the result of advances made in armament technology after the war in the field of high-speed diesels. Lightweight, economical and reliable, these engines were the standard equipment for ships propelled by fast diesels for many decades. Although impressive diesel units with very high performance and low weight were developed

(Below and following page) The PT.50 in section and below the deck plan of the main deck, The engine room was located exactly amidships to place the heavy engines exactly above the ship's centre of gravity. Like an aeroplane, all hydrofoils react very sensitively to shifts in the centre of gravity.

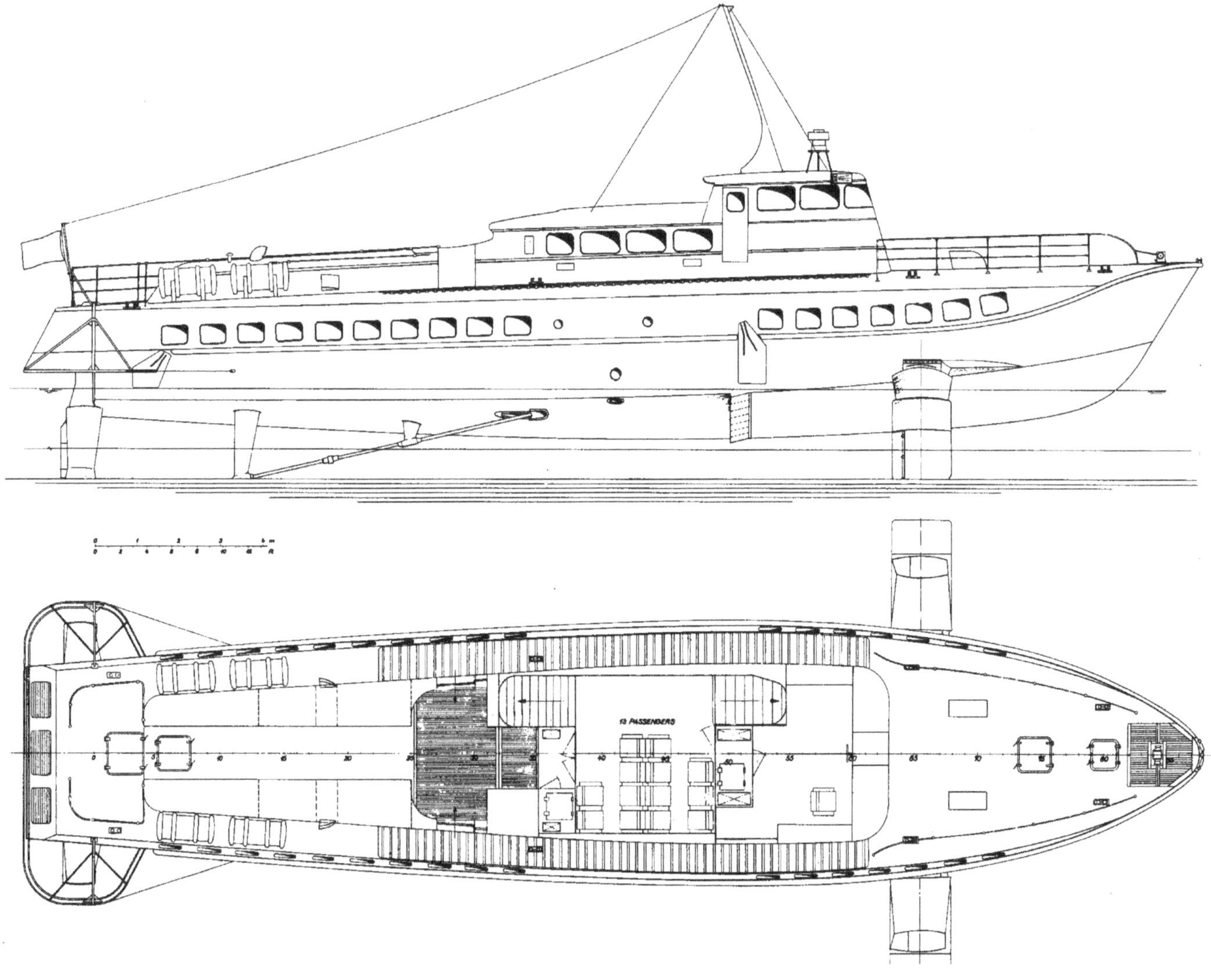

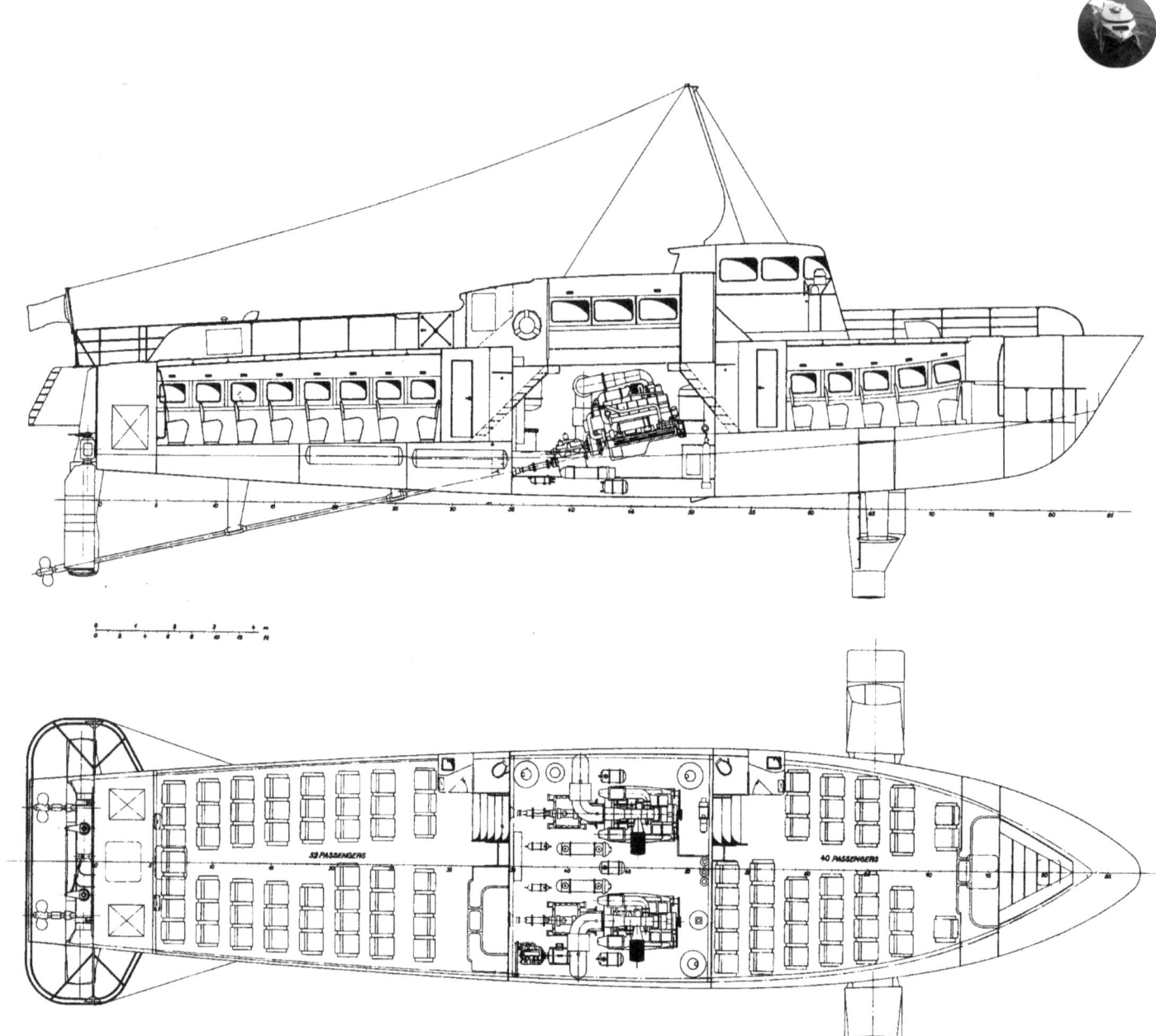

(Above) A drawing of the PT.50 - both the twin deck layout and the shape of the wings are clearly visible. This configuration has remained until the hydrofoils produced in Western Europe today. Only the size and passenger capacity have changed.

in the Soviet Union and in England in star form or as opposed-piston diesels, no other manufacturer came close to the reliability of the German machines as quickly. The experience with wing technology accumulated since the „PT.20" enabled the introduction of certain control mechanisms in the „PT.50". The wings were equipped with rudder surfaces, which at first enabled trimming and regulation of the „flight level" above the waves, but later were also actively used as dampers against the swell.

The „PT.50" was capable of carrying between 100 and 130 passengers. With a level of comfort that was not common at the time and an exhilarating speed of about 34 knots, she sailed across straits, between islands and along coastlines. In some places, the new hydrofoil robbed air transport of its raison d'être, such as on the route from Hong Kong to Macao. This connection was maintained by slow ferries and by flying boats until the introduction of the new racer.

The ferries survived, but they quickly became the mode of transport for people who could not afford tickets for the fast boats. Rumour has it that the hydrofoils also regularly smuggled gold from Hong Kong to Macau at the time. The high speed certainly made for a crossing unclouded by robberies.

decay never stops. There are only a few existing „PT.50s" left. One of the last was offered for sale in Italy around 2006 for a few hundred thousand EURO - the end of a career.

(Oben) Ein von Hitachi Zosen in Japan in Lizenz gebautes PT.50 in voller Fahrt. Es sind nur noch wenige Boote dieses klassischen Entwurfs übrig geblieben.

The „PT.50" was manufactured under licence by some shipyards, such as the Hitachi Zosen shipyard in Japan, where special laws made it difficult for foreign shipbuilding products to enter the market. This first successful phase of fast ferry shipping was 40 years ago. Many of the hydrofoils built then no longer exist.

In the meantime, enthusiasts of this technology are fighting hard for the survival of individual specimens. But

(Below) The PT 150 was the largest hydrofoil developed by Supramar. It operated between Copenhagen and Malmö in Sweden from 1968. In 1973, this service was discontinued and it experienced a chequered history.

The PT 150 was stripped of its wings in 1989 and then used as a small cruise ship. It is said to still exist.

Data of the original:

Length	37.9 m
Width with foils	16.0 m
Speed	32 kn
Passengers	250

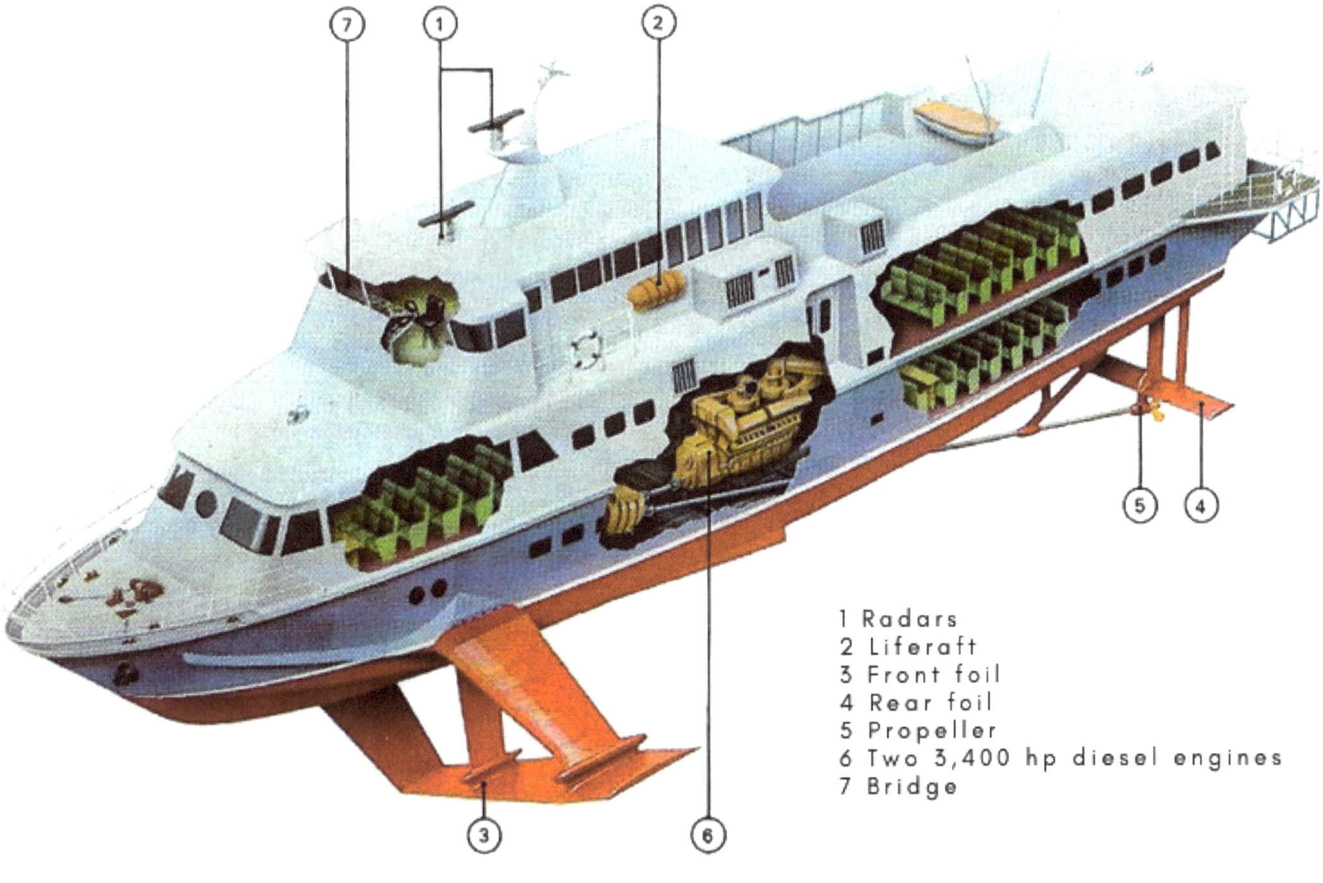

1 Radars
2 Liferaft
3 Front foil
4 Rear foil
5 Propeller
6 Two 3,400 hp diesel engines
7 Bridge

The Russian approach

After the conquest of Germany, the Soviet Union gained a precise insight into the state of development of v. Schertel's work. They captured some test equipment and many documents in Dessau in the Soviet occupation zone. In addition, the experts taken into custody there were allowed to participate „voluntarily" in the continuation of the work until the whole thing was transferred to the Soviet Union.

Today it is doubtful whether the Soviets really needed to fall back on looted knowledge when they had engineers like Rostislav Alekseev at their disposal. Born in 1919, Alekseev was trained as a shipbuilding expert in Gorky. In 1941, he began his first work on researching hydrofoil technology in the form of a detailed written paper. Although he initially had to interrupt his work due to the war because he became an engineer in a tank factory, he later received permission to continue the work from 1943.

The first test craft was quickly completed and gave enough positive results to continue the research with numerous other test boats until the early 1950s. Unlike other researchers abroad, Alekseev concentrated his work on developing hydrofoil systems suitable for the wide and shallow inland waters of the territorial Soviet Union.

This bore fruit, for on 25 August 1957, his first commercially viable boat, „Raketa 1", built at the request of the Ministry of River Navigation, sailed at over 60 km/h on the river route between Gorky and Kazan.

The robust and simple aluminium boat, with a length of

This illustration of a modern Russian „Kolkhida" can be used to explain the wing arrangement in the manner of Alekseev:
The front wing arrangement takes slightly more than 50 % of the total weight, but also provides the righting moment. The rear wing is equipped with the rudder and the propeller shafts and generates rather only lift with only little lateral stability.
The middle wing serves as a launching aid when releasing the hull from the water. Only a few people know that the suction effect of the water shooting forward at the fuselage during acceleration is so great that the fuselage can only be released with considerable effort. The problem is also well known in water aviation. Even small surface waves help considerably in detaching the fuselage from the water.

(Below) The Kolkhida in profile section.

Length:	34.5 m
Width incl. wings:	10.3 m
Draught floating:	3,5 m
Propulsion:	2 X MTU Diesel 12 V 396
	At 1,920 kW each
Speed:	34 kn
Passengers:	120 - 140

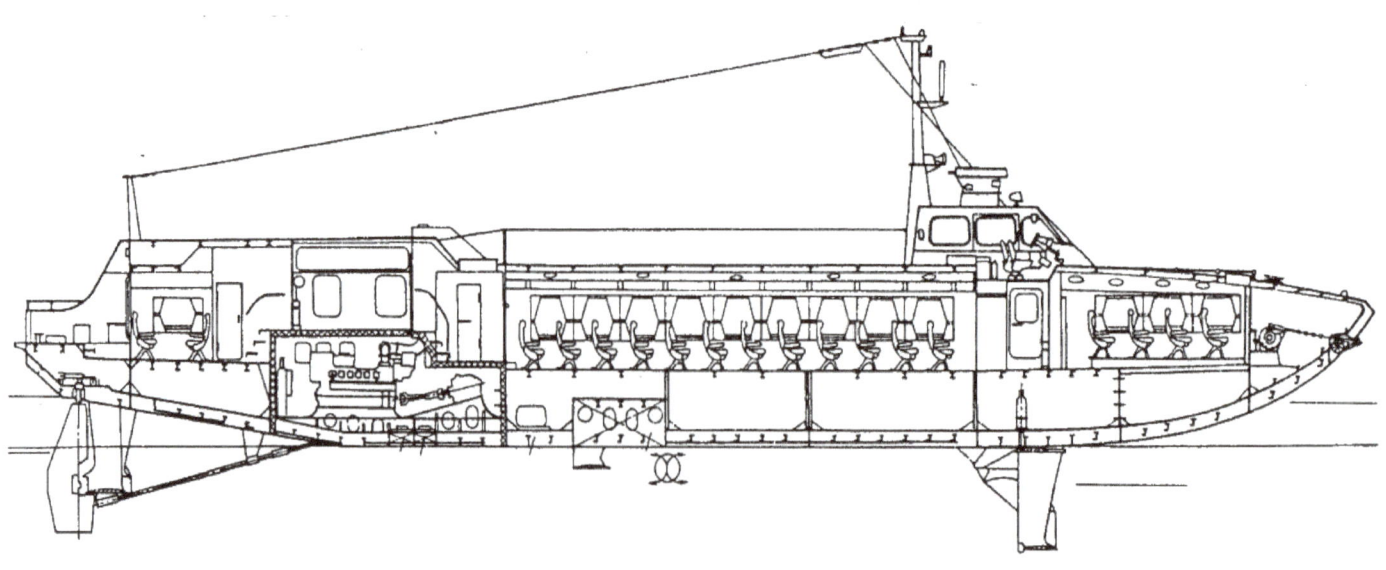

about 27 metres, could carry up to 66 passengers comfortably and quickly across the country's rivers and lakes. As it only had a draught of 1.80 metres, it was able to cross less deep places and thus became a universal river coach. The gigantic country, which began to rebuild itself economically and in terms of infrastructure after the World War, still did not have an adequate road network. Therefore, the rivers were an indispensable transport route for people and material.

As soon as the ice on the rivers melted in spring, the hundreds of Raketas and their successors were launched and raced along the rivers loaded with people, mail and supplies. The Raketa's robust and cheap diesel engines, taken from the T-34 tank, made a hell of a racket, but worked reliably for the most part. And if something did break down, they didn't spend much time on repairs. They simply replaced the defective engine and left the scrap metal where it was. In the planned economy, no-

The „Raketa" is one of the most manufactured hydrofoils in the world. Here, one sails past the Kremlin on the Moskva River.

Length:	27 m
Width:	5 m
Draught:	(floating) 5 m
	(on foils) 1.1 m
Passengers:	64 - 66
Speed:	33 kn
Propulsion:	One diesel engine of 900 or 1,000 hp

body asked whether something was economical. If something was produced, it was produced in such masses that spare parts were spare parts were available in abundance. Even today, you can buy large quantities of propellers, engines and so on in Russia at low prices, all of which were mass-produced without prior orders during the Soviet economy.

The valiant „Raketa" was soon followed by the larger series of „Meteoras" and „Kometas", which maintained links everywhere in inland and coastal areas of the country. Several hundred of each of these boat types were

manufactured. The exact number is not known. Alekseev's hydrofoil system differed considerably in its form and mode of operation from the V-wing system of Baron von Schertel.

Due to the shallow water depth of the rivers, the Soviet designer could not afford to be generous with the draught. He therefore developed a wing profile whose effect depended on the depth of water in which it was located. It was designed for a submergence depth that corresponded to its width (also called profile depth). If it exceeded this, the buoyancy increased completely intentionally to such an extent that the wing pushed the boat upwards again.

In addition, Alekseev ensured that the rear and front wings of the boat interacted with each other. He achieved this by giving both wings different angles of attack. If the front one dipped too much, the rear angle of attack also decreased even more. The rear wing produced less lift and the boat did not turn on its nose.

The above system was thus able to calmly restore the stable position over the larger front wing. This configuration of the wings has been preserved to this day in the boats built in the East and ensures that the boats sometimes sail a little bumpier in open seas than boats built according to the Supramar system.

The simple and compact design has been successful. About 80 percent of all hydrofoils ever built are technical descendants of the legendary „Raketa 1". Today, only the boats from the later „Voskhod" and „Katran/Kolkhida" series are still in use. They can still be used very economically. Moreover, the purchase prices for used examples are fantastically low and spare parts are available in large quantities.

One more thing about Rostislav Alekseev: he later turned to the development of the ground effect aircraft, known in Russia as the „Ekranoplan". His work in this field resulted in gigantic flying machines that chased over the surface of the water at speeds of over 200 knots at low altitudes.

Unlike the hydrofoils, however, this work failed to achieve ultimate success. Rostislav Alekseev died in 1980.

The „Meteor" became a standard vehicle for fast passenger river transport in many countries of the Eastern Bloc. It was not only efficient and robust, but also very shapely.
The pictures below show the typical carrying surface arrangements that are part of Alekseev's system.

(Above) The hydrofoil „Corsario Negro" built by Blohm & Voss in Hamburg was a licensed product of the American aircraft manufacturer Grumman. Due to considerable technical problems, this project remained unsuccessful. The only prototype was always a repair case....

(Below) The so-called „Project 80" from the Soviet Union in 1966 was an attempt to transport a tank wedged between two hydrofoils across the sea. In dictatorships, scientists who are popular with the authorities are often given too many resources ...

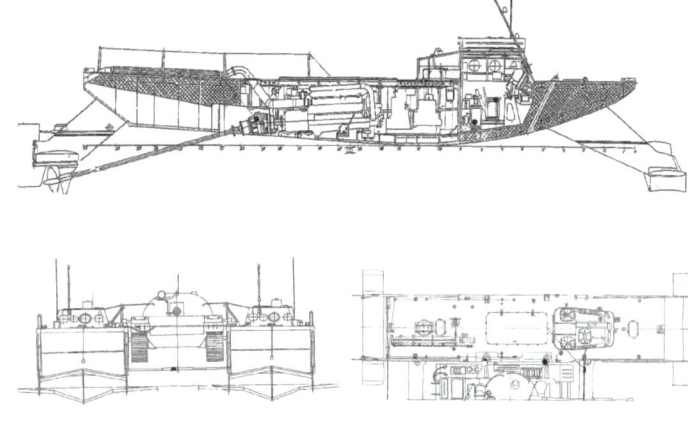

Interestingly, the US remained relatively unimpressed by the progress made in Europe and the Soviet Union. But they did not neglect the technology. Promoted by the US Navy, extensive attempts to make the hydrofoil militarily useful began as early as the early 1950s.

The US Navy was still under the strong impression left by the famous „PT-Boats", as the Navy's speedboats were called, in the fight against the Japanese in the island world of the Philippines. But everyone was aware of the weaknesses of these boats in terms of seaworthiness and range. Even in small waves, the speed had to be drastically reduced or the hull threatened to break apart.

Other branches of the armed forces, such as the US Marines, devised concepts for landing craft that could race from ship to shore at high speed. This was to help avoid casualties among troops before entering the country. Numerous experiments were made, such as with the flying amphibious truck.

The coasts of the USA have, in many sections, areas with sea states that far exceed those found in the Mediterranean, the Baltic or the Black Sea. Thus, wing systems that derived their stability and thus their seaworthiness from their shape alone were viewed with scepticism.

Experiments began with active controls of the immersion depths. These were initially purely mechanical, but with the advent and proliferation of the computer in the 1960s, it was possible to develop autopilots that could keep a hydrofoil level. Advances in aeronautical technology were implemented directly and without detours. Money for this was plentiful during the „Cold War".

The resulting technology of fully submerged hydrofoils produced the most powerful and seaworthy speedboats to date. But what does „fully submerged" actually mean?

This type of wing does not require a dihedral or specially designed airfoils. A wing is attached to a vertical stilt (American: „strut"), which has rudder surfaces like those of an aeroplane, which are moved by hydraulic or electric drives according to the commands of an autopilot. Not a piece of the wing protrudes from the water, so that the full surface can be used for generating lift. The wing can therefore be built comparatively small.

The autopilot is the actual heart of the system: it receives information from sensors about the distance of the hull from the water surface, about the speed, the position of the hull, acceleration forces acting on it and other data. From this, it determines the next required setting of the rudder surfaces in fractions of a second.

If everything works correctly, the boat can sail through the moving sea at a furious speed of 30 to 45 knots without anyone getting seasick or any other unpleasant situation occurring.

The main difference to the previously described designs is that a hydrofoil with fully submerged flights is no longer inherently stable, but if all the control systems fail, would immediately crash.

To prevent this from happening, all systems must have backup power units and the control system must run an emergency programme for rapid ditching should the autopilot fail.

Despite the high technical effort, this type of hydrofoil is faster, more seaworthy and also more economical with high payloads than the older inherently stable types. Their

The „Flying DUKW" from 1957, which was created from a wartime amphibious truck, was an adventurous experiment of the pioneering period after the Second World War. Although it could apparently go fast, it was not a success.

Here, too, faith in technology triumphed over common sense. The drive turbine from a helicopter consumed all the space for the possible payload. One often has to wonder a lot about the policy of allocating funds in armies.

disadvantage is that they are complicated and expensive to build. The extremely high proportion of electronic and hydraulic controls, most of which come from the aviation industry, exact their price.

The US Navy tested various concepts in the 1960s and 1970s before striking the big blow. The new class of ships known as „PHMs" was created and was to be deployed throughout NATO. However, the main NATO countries, such as the Federal Republic of Germany, were able to escape the grip of this programme and go their own ways. Thus, only six „PHMs" of the „Pegasus class" were built, which served in the Navy for a while. Stationed in Florida, they excelled in the hunt for cocaine smugglers.

smugglers, who for the first time found them to be a faster opponent. The story of a „PHM" that chased a smuggler equipped with a heavily motorised „cigarette" racing boat became famous among experts. The smuggler

propelled his boat to a dangerous 55 knots, but was still caught by the „PHM", which is said to have reached almost 65 knots at the time. It is reported that the „Chief" on board the „PHM" switched off all the overspeed fuses of the gas turbine. The operation of the „PHM" was very costly even without these show-stoppers. The fleet was decommissioned at the end of the 1980s and scrapped in 2000. Only the „USS Aries" has survived as a museum ship on the Mississippi. Efforts are underway to make this ship „airworthy" again. However, there is currently a lack of a cheap gas turbine for the main propulsion and, of course, of money. Since the PHM, no further relevant attempts have been made to build hydrofoil-based fighting boats. There are now better methods of getting hulls up to „speed". An important consideration is flexibility in use. A speedboat should always be usable in different speed ranges, not only very fast or very slow.

The US Navy's PHM, despite its impressive performance, could not fit into a globally operating fleet. Moreover, it was very expensive to procure. Thus, the six PHMs remained exotic racers that eventually only hunted drug smugglers. As of 2000, four were scrapped, another was preserved as a museum ship and the last was converted into a slow-moving yacht.

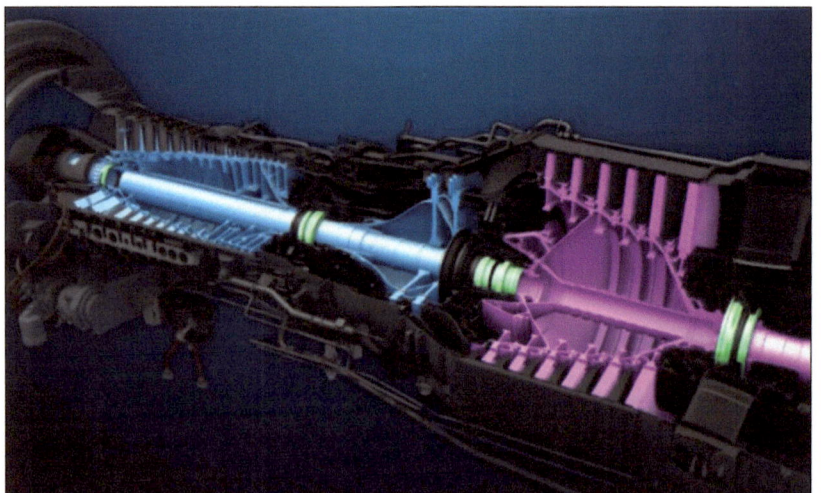

(Above) You can see here exactly why stilts are advantageous at sea. With them, the PHM can also sail over larger waves unimpressed.

(Above) The heart of a PHM - an LM 2500 gas turbine from General Electric. It provided the PHM with about 18,000 hp of power. Today's versions of the LM 2500 are even more powerful. (Right photo by Jan Huisman)

(Above) The Boeing Jetfoil is currently the end point of American hydrofoil development. Its performance and ability to operate in heavy seas has made it indispensable in some areas of the world, despite all the advances in non-"flying" fast ferries. It is currently being sporadically replicated in China.

Length:	27.4 m
Width:	8.53 m
Draught floating (hydrofoils lowered):	5,40 m
(hydrofoil raised)	2.20 m
Propulsion:	2 X gas turbines Allison 501 of 2,460 kW each
Speed:	43 kn
Passengers:	350-400

The technology used in the „PHM" could also be used in the civilian sector. The well-funded Boeing Corporation therefore embarked on a daring experiment to excel in civil shipbuilding. The result was the „Jetfoil", the first of which was launched on 29 March 1974. For a civilian ship, it combined an astonishing number of innovations. For the first time, the hull was welded together from marine aluminium. The hull offered space for a maximum of 350 seats on two decks, whereas 200 to 250 seats were normal. The passenger decks were modelled on the company's liner jets. The rear wing system, which carried the greater part of the load, consisted of three struts and a straight wing mounted between them. The middle strut was thicker than the others, and served as an intake pipe for the jet engine's water.

The front wing was attached in a T-shape to a single strut. All the wings could be folded upwards when the boat was travelling slowly, to allow it to access harbours with shallow water depths.

The propulsion system was driven by two aircraft gas turbines of 3,300 hp each, each driving its own water jet pump. These sucked the water in through the middle strut and, greatly accelerated, ejected it aft through two nozzles. This powerful drive enabled rapid acceleration to get the boat off the ground. When travelling slowly, the same drives were used in a heavily throttled form.

The very complex autopilot system of the „Jetfoil" kept the ship upright, provided controlled course changes and controlled the lift-off and watering of the hull. The helmsman therefore had no direct access to a rudder or similar, but passed on his commands to the computer via a control horn.

Thanks to the autopilot, the „Jetfoil" was able to operate at high speed even in sea states where previous speedboats had to give up. It covered its routes at a speed of about 43 knots in wave heights of up to 3.5 metres. This made the „Jetfoil" the predestined means of transport for areas where the swell was too high for other fast ferries. New connections were established on the Pacific between various Japanese islands and on the west coast of the USA. In Europe, various routes were established between the mainland and Great Britain. There was also a line with this type at the Canary Islands. However, Boeing had made a mistake in choosing aircraft gas turbines for propulsion. This was because the propulsion system required paraffin or at least well-purified light diesel oil as fuel, which was expensive compared to the usual gas oil.

Moreover, the maintenance of aircraft gas turbines is more expensive and more complicated than that of diesel

(Below) The cockpit of a Boeing Jetfoil. The yoke comes from Boeing's „Aviation" department. Joysticks have become common as controls on today's fast ferries.

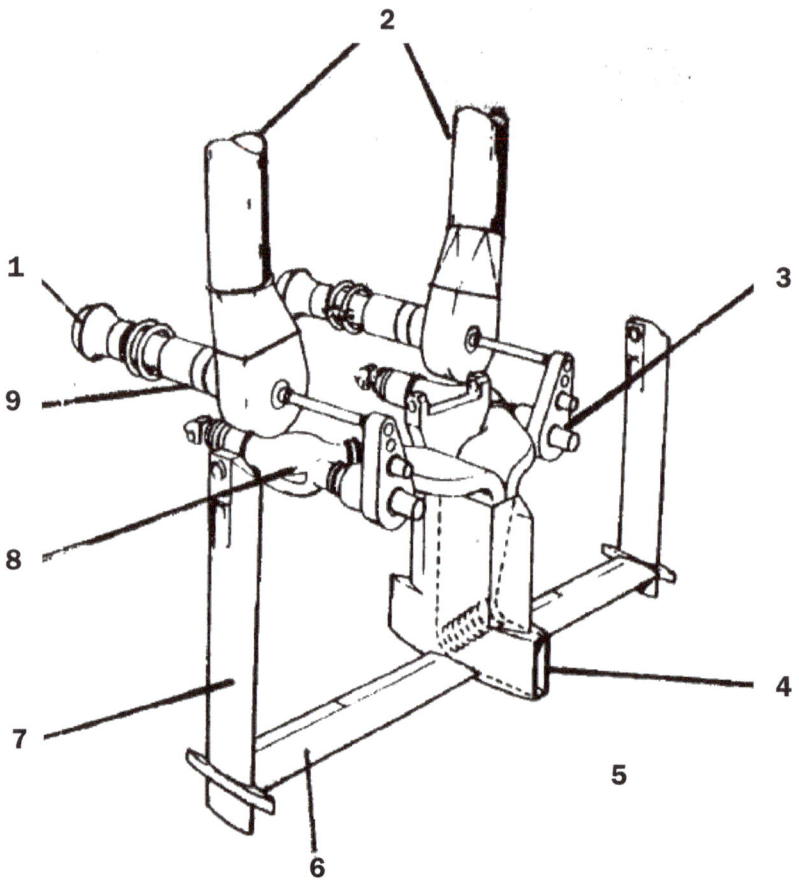

The propulsion system of the Boeing „Jet-foil":

1	Air intake of the gas turbines
2	Exhaust post
3	Reduction gear
4	Water inlet of the jet engine
5	Direction of travel (blue arrow)
6	Rear wing
7	Strut
8	Jet pump, water nozzle behind
9	Gas turbine

How it works:

The gas turbines drive both jet pumps, which suck in the water via the inlet opening (4), convey it through the struts to the pump and expel it again with great force at the tail. The recoil is so great that a „jetfoil" can accelerate very quickly.

engines. The not too great sales success caused Boeing to stop production after 15 copies. Later, more „Jetfoils" were produced in Japan by „Kawasaki Heavy Industries".

(Below) HMS „Speedy" was purchased by the Royal Navy in 1979 as a safety boat for offshore installations and fishing. As she could only run either very fast or very slow, it is not entirely clear how she could have performed this task. She was decommissioned in 1982. Today, after a conversion, she serves as a civilian fast ferry in Hong Kong.

Two jetfoils flying around Hong Kong.

She was the fastest of them all: The Canadian „Bras d'Or (FHE 400)" achieved the world record for unarmed military ships in 1969 with 63 knots.

(Above) The wings, made of hard stainless steel, are exposed to high loads. The paint is rubbed off after only a few trips. The wing of a „Kolkhida" can penetrate a wooden beam with a diameter of about 20 cm without further ado. But plastic foils that wrap around the wing profiles and lines that preferentially wrap around the propeller shafts are the real enemies.

The era of hydrofoils is definitely over. The great success in the 1950s and 1960s was followed by disillusionment over the fact that they had considerable weaknesses in daily use that prevented their ultimate lasting success.

In the meantime, new and easier-to-operate fast ferry types came onto the market, which quickly displaced the hydrofoil. But what were the disadvantages that led to its demise?

Hydrofoils, by their very nature, have „extremities" in the form of the wings and struts that are delicate and vulnerable. If a conventional small vessel runs aground on a sandbank, it is in most cases able to continue its journey after being towed free.

However, when this calamity befalls a hydrofoil, the effect is catastrophic in all cases.

Since the wings are usually severely damaged, a total loss of the boat usually occurs. Since the hull of the boat is at a certain „flying height" above the water surface, it hits the water with considerable force from that height during the crash. With the exception of the „Jetfoil", almost all hydrofoils require quite a large water depth at the moorings. For example, the „Kolkhida", which is about 34 metres long, can only call at harbours that have a water depth of at least 3.4 metres. Such restrictions considerably limit the application possibilities for these boats.

Last but not least, the maintenance of the boats is far too complicated for most shipyards. Especially the complex Boeing „jetfoil" makes great demands on the qualifications of the maintenance specialists. Much of the required knowledge about gas turbine maintenance and aircraft hydraulic systems is not part of the usual range of services offered by most shipyards.

In contrast, when the first catamarans from Norway and Australia came onto the market, they were found to be fast vessels whose simple design and engineering were better suited to the rigours of daily ferry service. Today, there are only a few hydrofoil lines left. Most of them are found on the coasts of Italy and Japan. In Italy it is the „Foilmaster" by Rodriquez and in Japan the licensed product „Kawasaki Jetfoil".

(Below) An experimental boat built in Germany with fully submerged wings. It did not yet have an automatic control system for stabilisation, but could sail quite well under manual control. Nevertheless, manual control by a jet pilot was not the desired solution. Today it stands in a southern German technology museum in Speyer.

(Above) If you pay attention to the details in this picture, you will notice that light shines through between the keel of the catamaran and the surface of the water. This fast ferry actually „flies"!

Hybrid catamarans - a cross between a hydrofoil and a catamaran - have only been built as one-offs in Japan, Denmark and Norway. This example was built by Fjellstrand AS in the early nineties. Although the concept was of excellent seaworthiness, it failed to convince potential customers because of the large draught when afloat and the gas turbine propulsion. The only one built is now sailing in Hong Kong. After all, it can reach up to 45 knots. The current status of this vessel is not known, as the era of hydrofoils is gradually coming to an end in Hong Kong as well. The reason for this is the structural age of the boats, most of which have been in service for more than 30 years, and cost pressures.

When the water police learned to fly

In the 1950s, the leadership of the Hessian water police came to the conclusion that the speed of the conventional police boats, which could only go a little faster than the river barges, would no longer be sufficient to follow the ever faster boats of the water sportsmen. Therefore, Supramar AG had a new police boat developed that could fly on hydrofoils. The new boat class, called „POT-3", was made of riveted aluminium and flew on steel wings. With their petrol engines, the boats reached speeds

The boats, with their petrol engines, reached speeds that were unusual for the river traffic of the time. The boats - three in number - served until the late 1970s without major problems. However, the engine noise and also the maintenance effort were too great compared to modern boat types to use them for much longer.

These Hessian police boats were the world's first successfully used non-military hydrofoils and came into service before the first PT. 10 on Lake Maggiore. What has become of the decommissioned boats, no one can say today. According to a former commissioner of the water police, one was scrapped, the others were sold at auctions and presumably deprived of their wings and used privately as normal boats. Rarely is such technology historically appreciated and preserved.

This craft have also been built for other countries, but these have not survived either.

Length:	*10.4 m*
Width (hull/beam):	*2.6/3.65 m*
Draught:	*1.2m at rest, 0.5m flying*
Mass:	*4.1 to*
Engine power:	*150 hp*
Maximum speed:	*30 kn*
Cruising speed:	*24 - 26 kn*

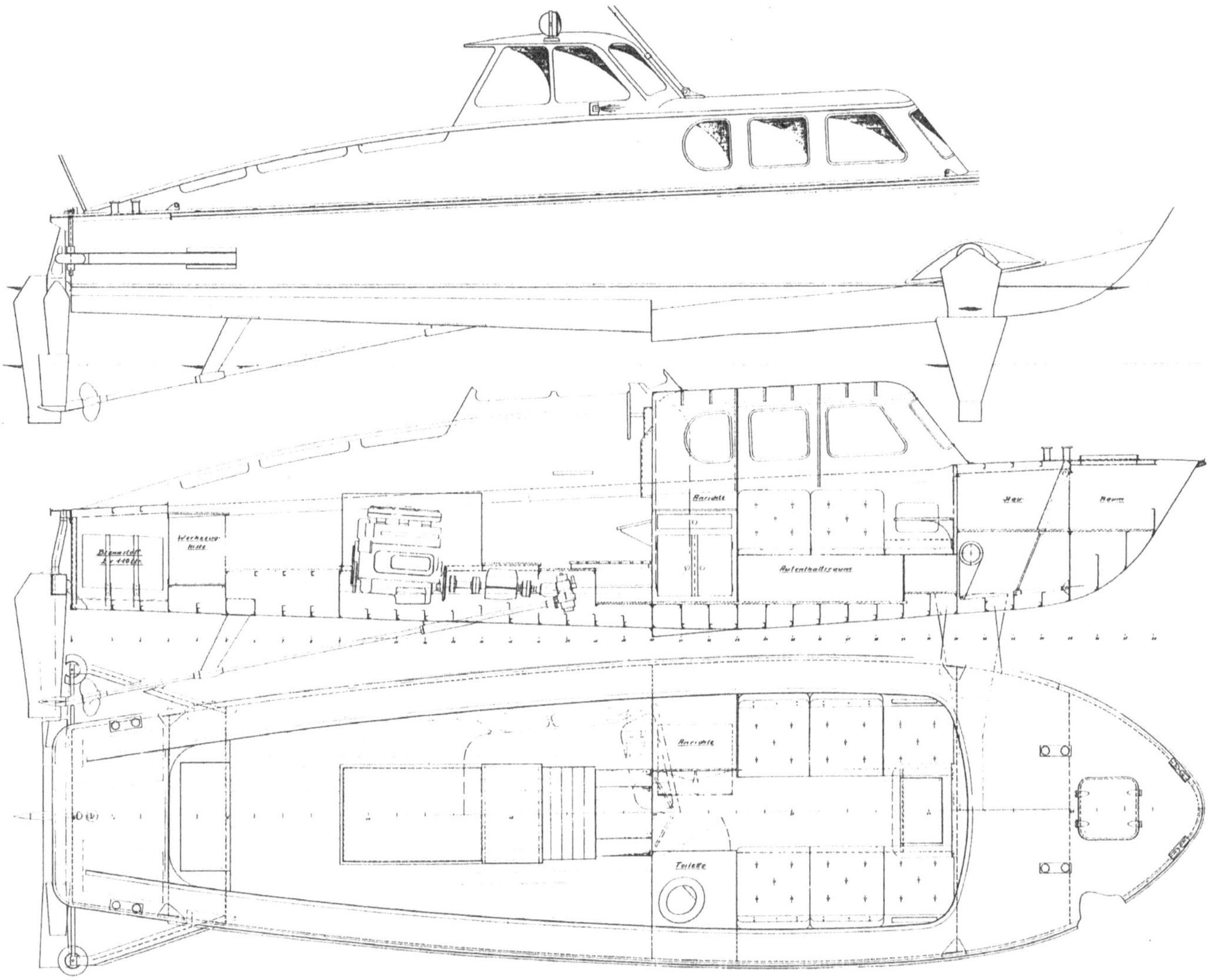

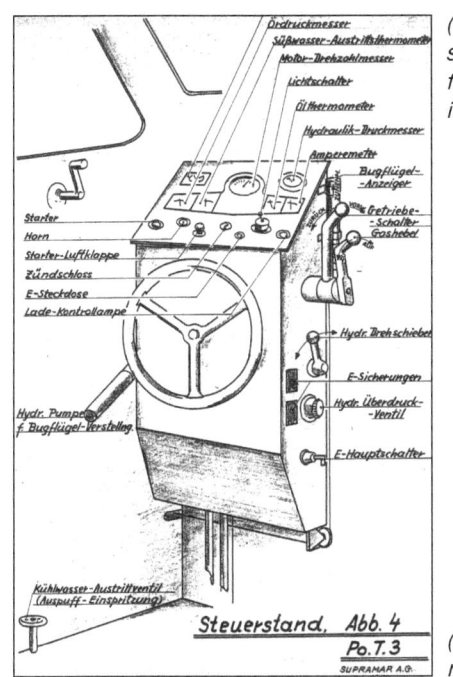

(Above) Although the hull of the POT-3 looks like that of a normal motor cruiser, there are considerable differences to a pleasure boat. The front wings could be swung all the way forward via a hydraulic hand pump. Therefore, the bearing of the wings on the hull is one of the strongest components. The hulls were riveted from aluminium sheets.

(Left) Drawing of the driver's stand in the POT-3, taken from the original operating instructions.

(Right) A POT-3 on lifted for reparis.

Powered only by the wind

Even though motorised hydrofoils have seen their demise, a completely different breed of flying boats are now racing the world's oceans.

The enormous progress made in recent decades has made it possible to build ever lighter and faster racing yachts under sail. At first, people tried out catamarans and trimarans in order to be able to sail faster and faster. But even though these boats could sail at speeds of up to 35 knots in stronger winds, their use became life-threatening in rough seas. Also, the world record sailors still demanded better performance.

As early as 1966, the American Dave Keiper built the first sailboat that could race along on hydrofoils. In 1970, he sailed the trimaran „Williwaw" (see picture above) at speeds of 15 to 20 knots from California to Hawaii. To do this, he used a ladder-shaped wing system that he made out of metal.

In comparison, only the legendary clippers of the 19th century could reach speeds of between 14 and very rarely 18 knots. But the effort required was considerable. Fast hydrofoil sailing, however, stalled in its development be-

(Above) Dave Keiper's „Williwaw" - the pioneer of modern racing yachts.

cause wind is a relatively weak form of propulsion compared to diesel engines and gas turbines. Therefore, flying sailboats had to be built extremely light.

Only plastics reinforced by carbon fibre and new types

A small board, a stick with a wing and a lot of skill are all that a kite-foiler needs. Modern carbon fibre reinforced plastics make it possible. (Photo by Bruno Sroka)

(Above) The L'Hydroptere in action. Her dimensions are enormous, yet she is built to be lightweight.

Length:	18.24 m
Width:	24 m
Displacement:	approx. 7 tons
Sail area:	400 m²

of stiff sail materials made it possible to build new types of hulls, super-strong masts and large sails with the properties of aircraft wings. In the 2000s, progress began to be made in the „foiler" scene. In parallel, someone invented the „foil-board". Basically, it was a surfboard with a wing attached to a handle underneath. There are versions for surfing, wind surfing and kite surfing. The „kite" stunt kite in particular, shaped like a parachute, is a powerful means of propulsion in strong winds. „Kite-foilers" reach enormous speeds and are just the thing for the adrenaline kick of adventurous water sports enthusiasts.

But back to the „big ones": today there are now a larger number of racing yachts that lift off completely with hydrofoils or prop themselves up in the water to be able to put even more sail area into the wind. The speeds are breathtaking: fully flying yachts can reach speeds of over of over 30 knots and lie like a board in the wind. This is because the lift of the submerged wings increases so rapidly with increasing submergence that the hull can be completely lifted again. The pure form lift of a hull would be much weaker.

This is why a monohull yacht with only one wing extended on the side facing away from the wind („leeward") can reach considerably higher speeds than a conventional boat. However, since it is also a full glider and its hull is only slightly immersed in the water, it needs all the help it can get to avoid capsizing anyway.

Sailing races with yachts supported by hydrofoils are a new and exciting form of sailing, regardless of the size of the boats - because there are also very small hydrofoil yawls.

In the heyday of yachting in the mid-19th century, the ancients would never have dreamed that today's racing yacht would race along for days at almost three times the speed of the fastest clipper.

By the way, the sailing world record for long distances is still held by a trimaran without wings. In August 2009, the 40-metre-long boat „Banque Populaire 5" sailed across the Atlantic from the west in only three days, 15 hours, 25 minutes and 48 seconds (!). The tour took place on the classic route of fast ocean liners between Ambrose Light near New York to Lizard Point at the southern tip of Cornwall. The average speed was 32.94 knots. This achieved a similar cruising speed using only wind power as the classic transatlantic ships such as the SS „United States" achieved using at least 210,000 horsepower. However, these racing yachts are also quite dangerous. If a bow tip dips a little too deep into a wave at the speed of about 60 km/h (about 32 knots), the boat sometimes rolls over immediately. The impact at this speed on the water or even on parts of the boat is then not only unpleasant for the crew, but life-threatening. Water at this speed is very unforgiving.

(Below) The design of the L'Hydroptere is more reminiscent of an aeroplane. In fact, it is.

(Above) But there are also small foilers. This is a 3.55 m long boat of the international Moth class. It can also reach high speeds of over 25 knots and is sailed by one person.

One of the most spectacular racing yachts was the huge French „L'Hydroptere", which in 2009 set an official world record of 50.36 knots (90.82 km/h) over a distance of one nautical mile. But she could also cover long distances quickly. On 9 February 2005, she crossed the English Channel with an average speed of 33.3 knots. Incidentally, the wind speed for the short distance record was only about 30 knots. Her performance potential was enormous. In another record attempt, she reached 61 knots (113 km/h)! That is a speed that most motorised

(Above) In pursuit of a world record. The speed corresponds to that of a car on a country road.

hydrofoils have never reached. Unfortunately, she overturned during this trip and had to be extensively repaired. A world record trip to circumnavigate the globe in under 40 days had to be abandoned. The massive racing boat was sold in 2017.

L'Hydroptere's top performance has since been beaten by another hydrofoil sailboat. This is the „Vestas Sailrocket 2", a strange-looking little boat that can only carry one crew member.

The performance of these boats, which are completely new in seafaring, show that the „future of sailing racing" belongs, at least in part, to „foilers". Increasingly lighter solid materials such as carbon fibre reinforced sandwich material, stiffer and lighter sails and other innovations will enable even more amazing performances.

It seems strange why a sailboat can reach several times

that speed in a 30-knot wind. It used to be the other way round, because the wind blew faster than the sailing boat could go. Now - the vectors of the wind and the airstream generated by the boat make this possible. The boat braces itself against drifting sideways by the lateral resistance of its wings or its hull flanks. At the same time, it sails in an optimal direction to the wind so that it can add the power of its own airstream. Good sail tacticians calculate this in advance and thus optimise the performance in the race. The extremely light racing boat, however, has only a low resistance and can therefore make optimal use of these power ratios. Presumably, many more exciting advances can be expected in this area of sailing.

(Above) Two foil racing yachts race each other in the America's Cup - The speed is around 30 knots.

(Left) This racing yacht of ‚Team New Zealand' is a 75" yacht for the America's Cup. Even as a monohull, she can reach enormous speeds of over 30 knots. The wings provide stabilisation against the pressure of the wind. Therefore, the „leg" is raised to wind-ward (the side facing the wind), as it is not needed. Steering such racing boats is very difficult and almost an acrobatics.

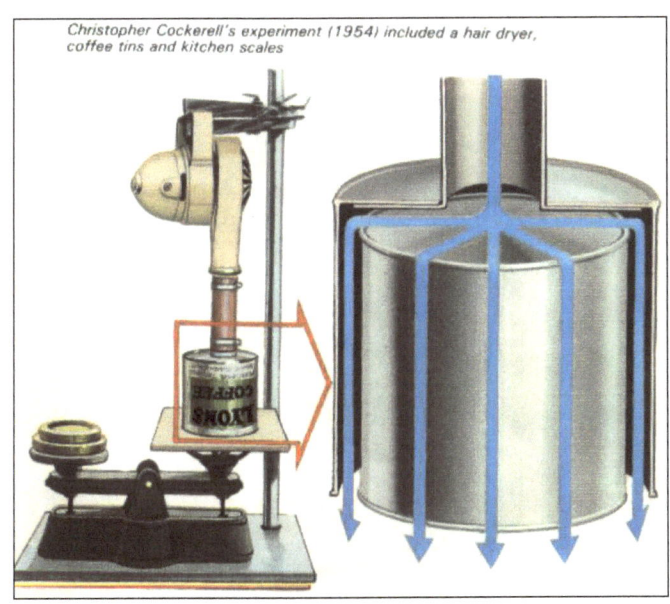

Christopher Cockerell's experiment (1954) included a hair dryer, coffee tins and kitchen scales

A fundamentally different strategy to break the resistance of the water is used in air-cushion vehicles. The direct contact of the hull with the water is prevented by inserting a separating layer of air. The resistance is thus so low that fantastically high speeds are possible.

The appearance of large hovercrafts seems to have as fascinating an effect on most observers as the sight of a space shuttle taking off. In fact, both devices have something in common, they move with great violence

(Right) Christopher Cockerell's first experimental setup. With simple means, he was able to demonstrate the higher efficiency of an air cushion operated at the edge by a kind of nozzle compared to a simple filled air chamber. The experimental setup can still be seen in a museum today.

and noise. In contrast to the quiet elegant hydrofoils or the sober catamarans, the spectacularly appearing „SR.N 4" canal ferries with their extreme noise emission have always been admired unreservedly - by laymen. The experts, however, knew quite quickly that a good show does not necessarily guarantee profitability.

Since the experiments of the engineer Roger Thorneycroft with small models around the end of the 19th century, a lot of water had flowed down the Thames before the Briton Christopher Cockerell took the first systematic measurements of the load-bearing capacity of an air cushion with the help of a powerful hairdryer, several tin cans and a pair of scales.

He discovered that not only did filling a cavity above a surface with air from a blower produce the best results, but that creating a curtain of air at the edge of the device to be lifted greatly improved the effect. The original experimental device of Sir Christopher Cockerell, now deceased, can be seen today in the „British Hovercraft

The principle of the hovercraft is simple:

A fan (1) pumps air into the fuselage (2), where it is distributed to various outlets. One part is blown directly into the air cushion (3), the other part flows into the skirt (4), which can be recognised from the outside by the rubber bulge. From there it continues into the apron fingers (5). They direct the air in such a way that a kind of curtain is created to hold the air cushion together. A blow-out gap forms between the bottom and the underside of the fingers.

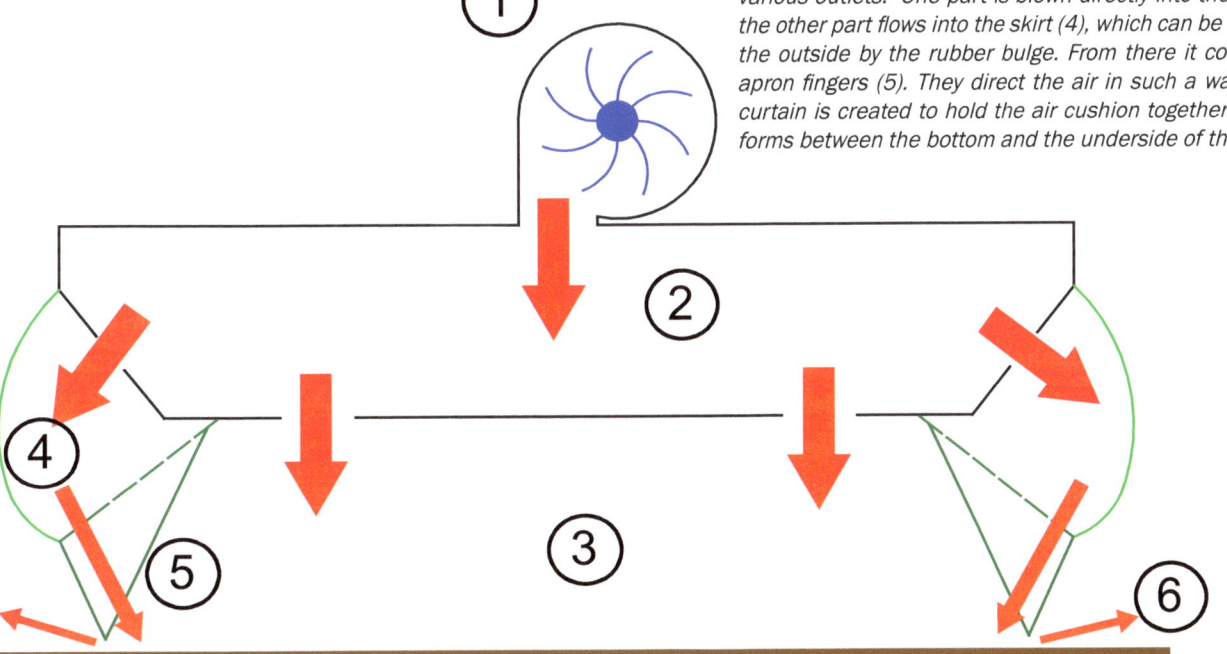

(Above) With the SR.N-4, the young British hovercraft industry landed its big blow that made the world sit up and take notice. Weighing almost 200 tonnes, the vehicle rushed from channel coast to channel coast at the unusual speed of 50 knots in the early 1970s. To this day, it remains one of the most fascinating means of transport ever developed by man. However, it is also one of the noisiest.

Museum" in Portsmouth. The first experimental devices were soon followed by functional models soon followed, chasing over flat places at a remarkable speed. After years of pleading with the British Ministry of Aviation, trials with larger vehicles were finally approved. The aircraft manufacturer Saunders-Roe built the experimental „SR.N 1" in 1959, which crossed the English Channel from Calais to Dover at high speed on 25 July of the same year.

The success was grist to Sir Christopher's mill. But something crucial was still missing: the hull of the „SR. N 1" floated safely, but only at a low height above the ground. Overcoming waves or obstacles higher than this hovering height of about five to ten centimetres quickly became dangerous or even impossible.

The Englishman C.H. Latimer-Needham had invented a flexible sheathing for the edge of the air cushion at the same time. He sold the patent in 1961 to Westland, the parent company of Saunders-Roe. The invention had found a way to insert a flexible „skirt" between the hard fuselage and the air curtain on the ground, which buckled when it hit an obstacle and allowed the vehicle to adjust

to the level of the obstacle. This invention made the „hovercraft" an all-rounder for off-road driving.

A skirt, for example, on a 25-tonne vehicle about 19 metres long, consists of air bags made of 1 - 2 mm thick neoprene with nylon reinforcement. Underneath are the fingers, which can easily be replaced when they become worn.

The hovercraft really floats on a flat smooth surface at a height of one to two centimetres. Nevertheless, the air pressure in the cushion is low and the ground is subjected to very little stress. Hovercraft are the only vehicles that can even negotiate rapids without danger, because they are not touched by the force of the water.

The fingers and the other skirt parts wear out quickly. This makes the vehicles very expensive to operate, in addition to the high fuel consumption.

(Above) The world's first long-distance hovercraft, the SR.N1. Its inventor Cockerell used it to cross the English Channel from Calais to Dover on 25 July 1959 - the anniversary of the first Channel crossing by aircraft. The journey took two hours.
The later hover ferries did the job in under 40 minutes with much more fanfare. This hover is now in a museum.

(Below and above) Hovercrafts don't have to be giants. The CushionCraft CC7 was developed in the late 1960s and sold in small numbers to Africa. It was the first hovercraft powered by a gas turbine,
Instead of being propelled by propellers, it was propelled up and over by two centrifugal fans. Deflection flaps made it very manoeuvrable. It could reach about 50 knots.

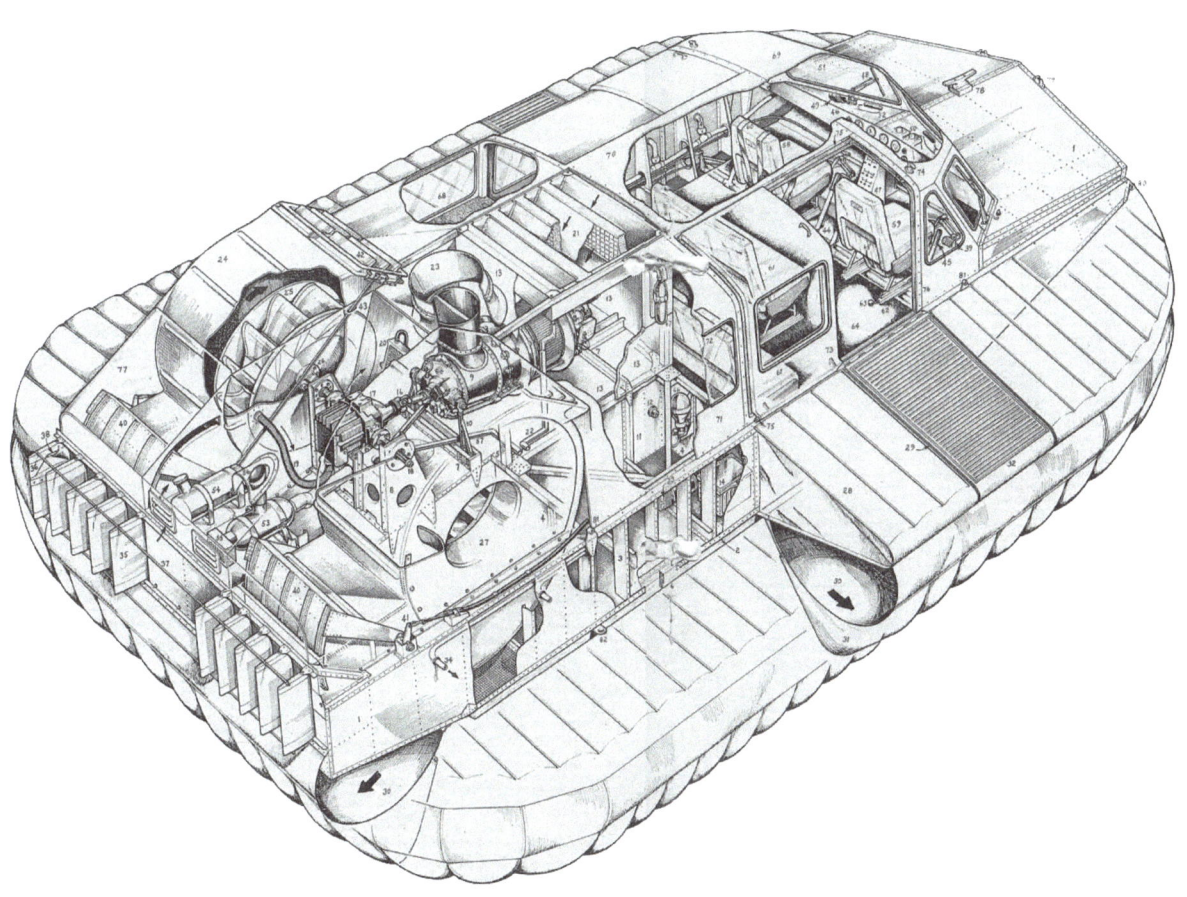

The „SR.N 4" not only completed impressive transport feats, but also had the honour of being one of the noisiest means of transport ever invented by man, together with Concorde and some moon rockets.

Nevertheless, the four gigantic propulsion propellers on the roof of the vehicle and the four propulsion gas turbines were allowed to rumble completely undamped for decades. The noise was so great that the „SR.N 4" could often be heard at the arrival terminal when she had only covered half the distance across the English Channel.

However, she proved to be weatherproof, reliable and safe. There was only one serious accident in all her years of operation. The commanders of the craft often said that they felt safer travelling at 50 knots on the English Chan-

nel than on a slow ship. The hovercrafts were not only very fast, but also very manoeuvrable. All slower vehicles in the intense main traffic in the strait were considered „standing obstacles", so one simply curved between them. In addition, the hovercrafts crossed the sandbanks in the middle of the Dover Strait, which were closed to shipping traffic, at high speed without being challenged. No one was travelling there anyway.

The trips were only suspended when the wave heights reached 2.5 to 3 metres. But until then the „SR.N 4" could

could be used unhindered at a flight height of 2.7 metres (distance between ground and hull). The terminals consisted of flat beach sections and concrete areas for parking

(Right, centre and bottom) Impressions from the SR.N-4 (Super 4): Passenger compartment, vehicle deck and cockpit

The SR.N. 4 Super Four:

Length:	56.38 m
Width (without skirt):	23.16 m
Propulsion:	4 gas turbines of 3,800 hp each
Passengers:	418
Cars:	54-60
Take-off mass::	300 to
Speeds:	
Maximum:	65 kn
Cruising speed: approx.	45 -50 kn

the hovercrafts. Dredged ferry ports were not necessary because the hovercraft ferries were completely amphibious.Six „SR.N 4" were built, two of which belonged to the longer and, weighing 300 tonnes, the heavier „Super-4" version. These were also the last to be used when they were decommissioned in 2001.

As no buyer could be found, they were scrapped two years later. Although the travelling public was very fond of these ferries - as their use considerably shortened the journey time to or from England - fate was not particularly kind to them. The oil price shock of 1973, the subsequent cost increases and the environmental impact of noise made their economic survival very difficult.

Their main problem was that far too much paraffin had to be used to carry a fairly small number of passengers and cars quickly. Against catamaran ferries, which were only about 10 knots slower and had many times their capacity, these „floats" no longer had a chance. Thus, in the early 1980s, the operators' decision to rely on large catamarans from Australia in the future put an end to hovercraft traffic.

However, the technology was also very costly: Every „SR.N 4" hovercraft had to undergo an all-round inspection of the skirt at night. Thus, repairs worth several tens of thousands of pounds were carried out on many nights. Operating costs almost completely ate up the income in almost every year. No real money was ever actually made with these vehicles. But they were a showpiece as a British invention - another similarity to Concorde.

53

Besides the gigantic dinosaurs of the hovercraft scene, vehicles in smaller size categories have evolved more inconspicuously and much more successfully. In Britain in the 1970s, it was quickly realised that propulsion by gas turbines had several disadvantages: for one thing, only trained aircraft technicians could maintain these engines, and for another, the specific fuel consumption was simply too great.

The appearance of compact air-cooled diesel engines enabled the „British Hovercraft Corporation", which had emerged from the Saunders-Roe Hovercraft Division as a subsidiary of the Westland Group, to construct the 21-metre-long „AP 1-88". This set new standards.

The „AP 1-88" was welded together from aluminium in the manner of shipbuilding and, powered by four diesel instead of gas turbines, was in no way inferior to its predecessors in terms of performance.

It could reach up to 45 knots with 82 passengers on board, but fuel consumption was many times lower than the older turbo monsters. The noise emission no longer corresponded to that of a jet, but only reached the noise level of a truck due to the shrouded propellers and the mufflers of the engines.

Thus, the medium-sized hovercraft now became gener-ally socially acceptable and was enthusiastically ordered in larger numbers. The „AP 1-88" has been used not only for passenger transport, but also as a military vehicle, as a transporter for bulky goods and as a drilling rig supply vehicle.

Whether in the Arctic, in Siberia, on the Caspian Sea, on Canada's rivers or in Africa - as a robust all-purpose vehicle it was up to almost any task.

In contrast to its predecessors, most AP 1-88s were not used as passenger ferries, but as a sober working tool - a floating truck. A good example is provided by the „Waban-Aki", an „AP 1-88" operated by the Canadian Coast Guard.

Instead of a passenger compartment, it has only a small deckhouse and a loading area on which equipment of all kinds can be stowed. It is used in winter for island resupply, ice breaking and rescue. In the summer, it helps with the removal of environmental damage and numerous other missions. Even today, „AP 1-88" are still sought-after items on the market, as they are no longer built.

The AP 1-88 has become a true „workhorse" of the hovercraft industry. Applications worldwide range from tourist transport and oil well site supply to water control and rescue. A robust structure, four reliable diesel engines and high performance have been the secret of success.

Length:	21.3 m
Width:	11.0 m
Total weight:	33.5 to
Payload:	7.2 to
Passengers:	up to 90
Propulsion:	2 diesel 390 hp
Air cushion generation:	2 diesels 390 hp each
Speed:	up to 50 kn

(Right) The larger successor to the AP 1-88 is the BHT-130, which is currently the largest diesel-powered hovercraft in the world.

Length	29.3 m
Width (incl. skirts):	15m
Weight:	75 to
Payload:	20 to
Passengers:	131
Speed:	45 kn

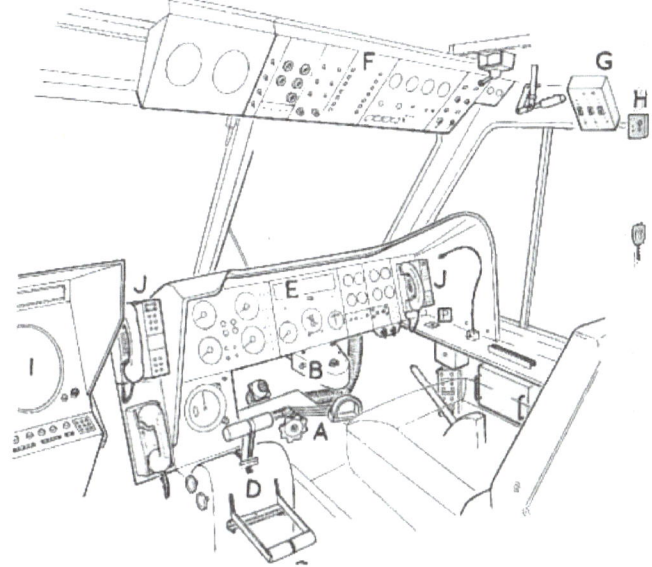

(Above) Drawing of the AP 1-88 cockpit

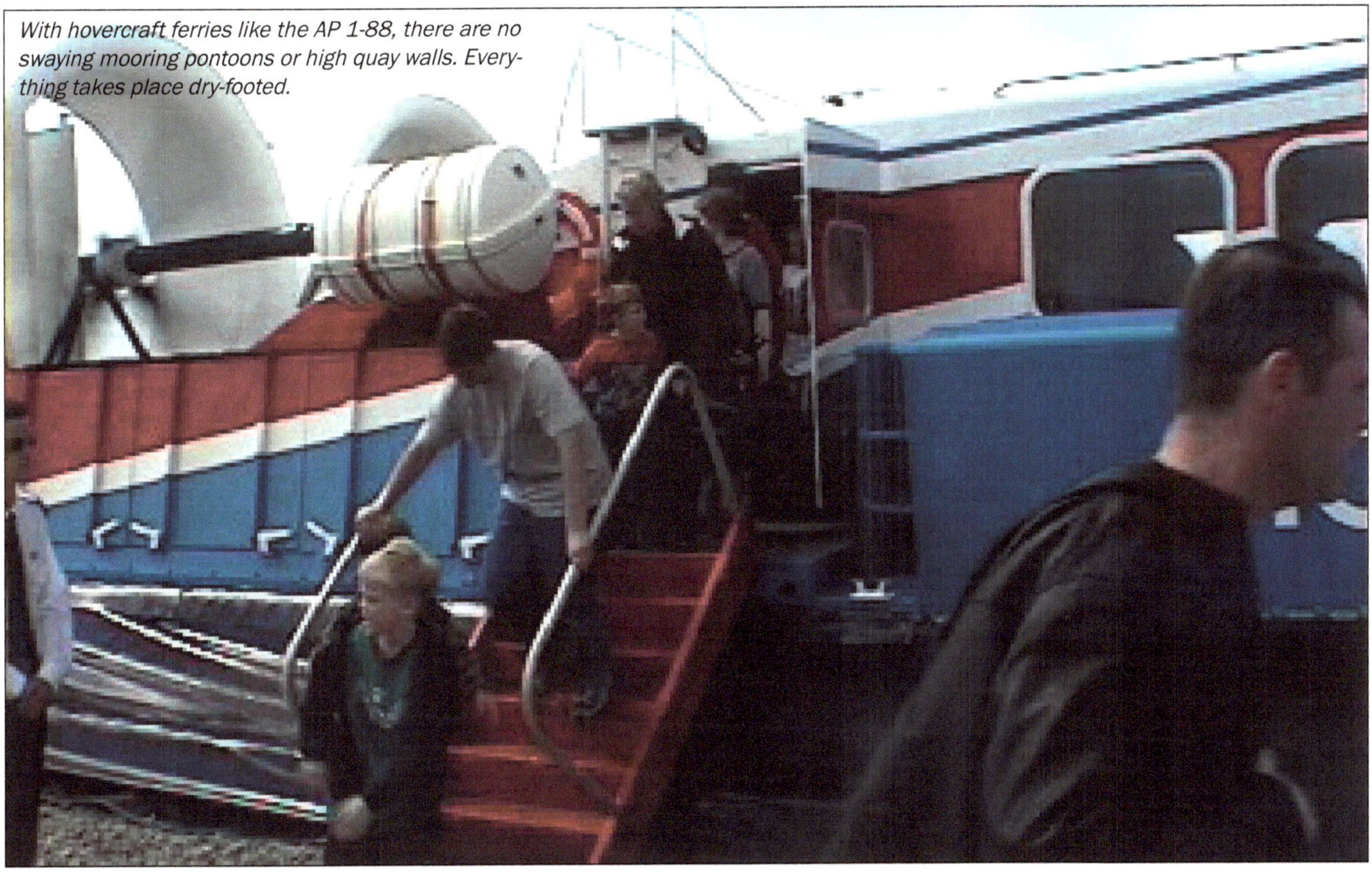

With hovercraft ferries like the AP 1-88, there are no swaying mooring pontoons or high quay walls. Everything takes place dry-footed.

Despite the simple basic principle, a modern hovercraft like the AP 1-88 is a complex vehicle.

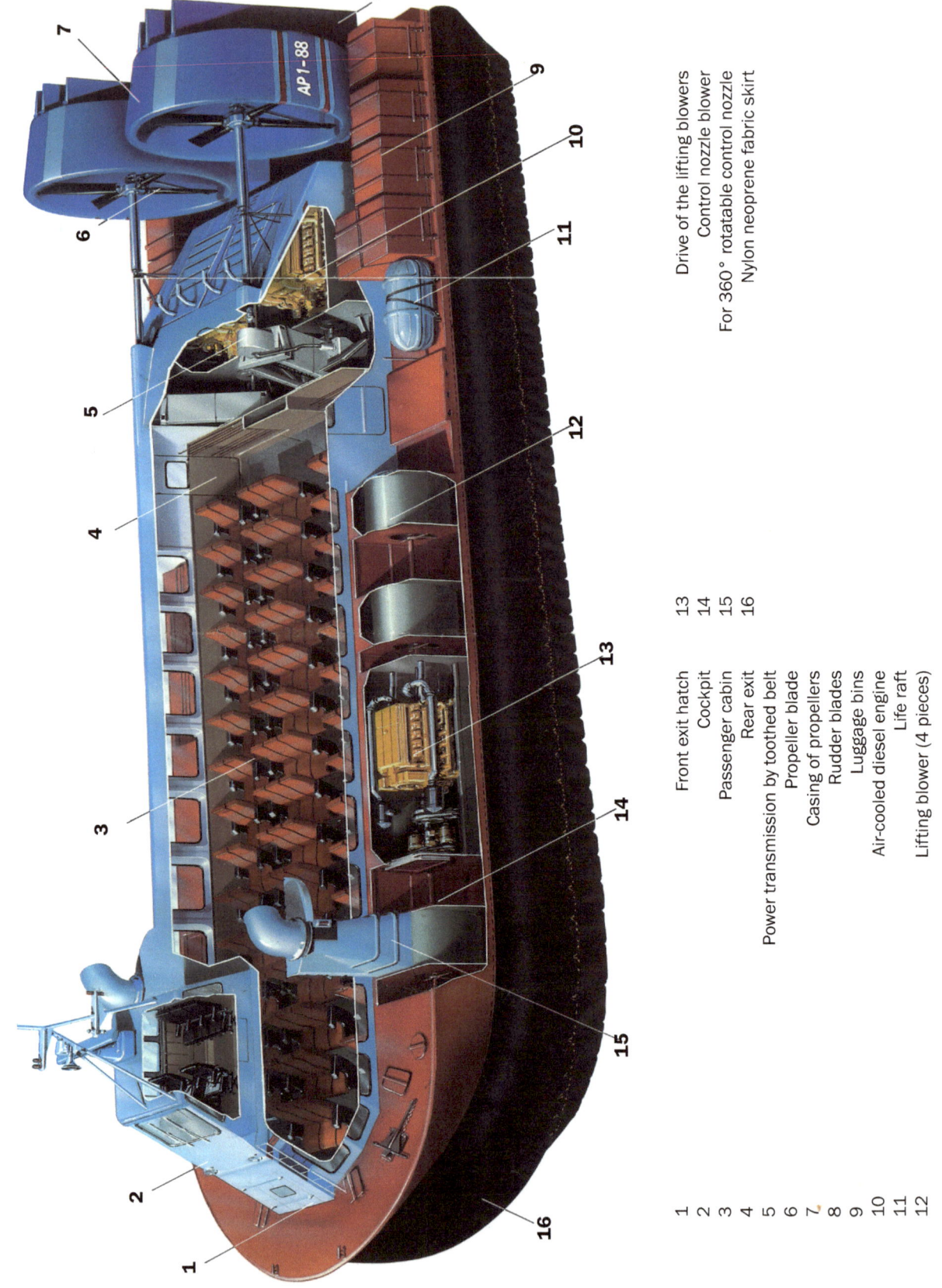

1   Front exit hatch
2   Cockpit
3   Passenger cabin
4   Rear exit
5   Power transmission by toothed belt
6   Propeller blade
7   Casing of propellers
8   Rudder blades
9   Luggage bins
10 Air-cooled diesel engine
11 Life raft
12 Lifting blower (4 pieces)
13 Drive of the lifting blowers
14 Control nozzle blower
15 For 360° rotatable control nozzle
16 Nylon neoprene fabric skirt

A classic hovercraft

The SR.N 6, together with its somewhat smaller „brother" SR.N 5, was one of the most widely used hovercrafts. Whether as a ferry, a rescue vehicle or a gunboat on the Mekong, both symbolised the variety of possibilities that could arise from the expanding technology in the mid-1960s.

Propulsion was provided by a gas turbine, which drove a single propeller and a radial fan for the air cushion. The cabin was elongated and had its access at the front. The pilot sat behind the windscreen on the right-hand side These hovercrafts underwent numerous modifications - in keeping with the experimental spirit of the time.

Several different apron systems were tried before a satisfactory configuration was arrived at with the „bag-finger" („bead and finger") combination. The SR.N 5/6 was strongly powered and could reach up to 60 knots in various cases. However, it took some time to get it to not roll over in high waves by using an appropriate skirt design.

These vehicles were tested in all sorts of places around the world. One even covered several hundred kilometres in sand in the Sahara, proving that hovercrafts can be very useful there as well. The SR.N 5 was produced in a number of 14 pieces. In addition, Bell Aerospace in the USA built a few more for various users in the States. The best known of these were the hovercrafts used by the US Army in the river deltas of Vietnam against the Vietcong.

The SR.N 5 and 6 were even real best-sellers: 57 units were sold from England to government agencies, armed forces and some civilian users worldwide, and more were built under licence.

It was not an optimal vehicle for ferry operations, as it had too little seating capacity. But as an all-purpose util-

(Above) Somewhat squeezed behind the dashboard sits this „hover pilot" of an SR.N 6. Moving the control stick back and forth controls the pitch of the propulsion propeller. Pedals move the rudders and a throttle lever regulates the hover height via the turbine speed.

ity vehicle it was unsurpassable, rivaling transport helicopters in particular. The SR.N 6, along with the large hovercraft ferries from the English Channel, contributed most to the public image of hovercrafts, as seen in many feature films.

Unfortunately, the publicity was not always in the hovercraft's favour, as it was a very noisy craft due to its open propeller and undamped gas turbine. Nevertheless, after more than 30 years, some SR.N 6s are still in active service, for example with the Canadian Coast Guard.

However, it has to be said that hovercrafts have often been overestimated by those who believe in technology. They are by no means the all-purpose amphibians they have often been thought to be. Especially in Vietnam, the US forces had to learn that driving a hovercraft through unfamiliar terrain can be a fatal mistake. A protruding

(Left) The SR.N 6 was developed by aircraft engineers and is therefore a correspondingly complex device. Compared to today's hovercrafts, it is more difficult to maintain and also requires expensive and rare spare parts.

Length:	14.8 m
Width:	7.7 m
Total weight:	10 to
Payload:	3 to
Passengers:	38
Propulsion:	1 gas turbine 900 hp
Speed:	52 kn

tree stump, a tank barrier or even military barbed wire can quickly destroy the skirt holding the hovercraft together and there goes the hovering.

Hovercraft are lightly built. They are not armoured and can easily be riddled with an infantry rifle.

Today, hovercrafts are mostly used in the oil and gas industry where mud, ice or snow obstruct normal trucks. But they consume an enormous amount of fuel and have a low payload. Their use requires very well trained personnel: hovercraft pilots form a very small and exclusive group in the community of vehicle drivers.

The author once looked after a medium-sized transport vehicle of this type for two years. Even small repairs are difficult, as special spare parts and a lot of expert knowledge are often needed. Nevertheless, they are fascinating and exotic devices.

Griffonhoverworks - as the company is now called - built a larger number of the very practical 2000 TD. The 12.7-metre long and 6.2-metre wide vehicle can reach about 34 knots and is powered by a 355 hp air-cooled Deutz diesel engine. It is demountable and can be easily carried on a truck or by a C-130 Hercules. This example patrolled the port of Basra for the Royal Navy after the Second Gulf War.

In the late 1990s, Albert Blum, a wealthy entrepreneur from Switzerland, came to the conclusion that the use of fibre composites, which had been common in yacht and aircraft construction for some time, would contribute significantly to improving the performance of hovercrafts. He first had a team of engineers develop and build a two-seater test carrier, the so-called „DONAR". Powered by a BMW 12-cylinder petrol engine, this vehicle was capable of developing a speed of 120 knots on a flat surface. The hull and the interior were characterised by a shapely design. It looked like one of the futuristic prototype cars that are regularly presented as design studies at the Geneva Motor Show. The DONAR proved the superiority of plastic in terms of weight savings and robustness. So they set their sights on bigger targets.

With the cooperation of various prominent experts, the ABS M-10 was created under the direction of the company ABS Hovercraft Ltd. in Great Britain, which was presented to the public in 1998. It was about 20 metres long and could carry either 90 passengers or up to 10 tonnes of cargo with a total mass of 40 tonnes. The passenger variant is known as the P-92. This vehicle, assembled from glass-fibre reinforced plastics, had a number of innovative features to solve those problems that had repeatedly arisen in the use of the predecessor vehicles.

For example, the drive was housed in an encapsulated engine room that was shielded against spray water and dust. This significantly curbed corrosion on the drives and electrical systems. The salt water mist that hovercrafts sprinkle around them is notorious for its ability to penetrate the smallest cracks and promote decay there.

In any case, only two diesel engines were needed instead of three or four compared to other hovercrafts of the same size. The weight savings from the plastic allowed the M-10 to carry about 30 per cent more cargo than older hovercrafts of that size.

The innovative product was rewarded with a high industry award from the UK government in 1998. Market interest was so great that the navies of Sweden and the island of Sri Lanka each ordered a vehicle for military transport operations. The Swedish Navy, however, assembled its M-10 itself, somewhat modified, at Kockums Verft in Sweden, while the Sri Lankan Navy had itself supplied from England. The slightly smaller prototype was later sold to the Belgium-based survey company EUROSENSE. Unfortunately, the initial successes were not followed up by follow-up orders, so the programme faded away for the time being.

Although this hovercraft was extremely innovative, the interest of the potential civil and the military operators was very few. A reason may be thaht hovercraft are very specialized and do not fit on any purpose.
At the time, the M-10 appeared on the market the time of the larger hovercrafts have been over and they became a nishe product.
The infrastructure is world wide not really made for amphibious operations.

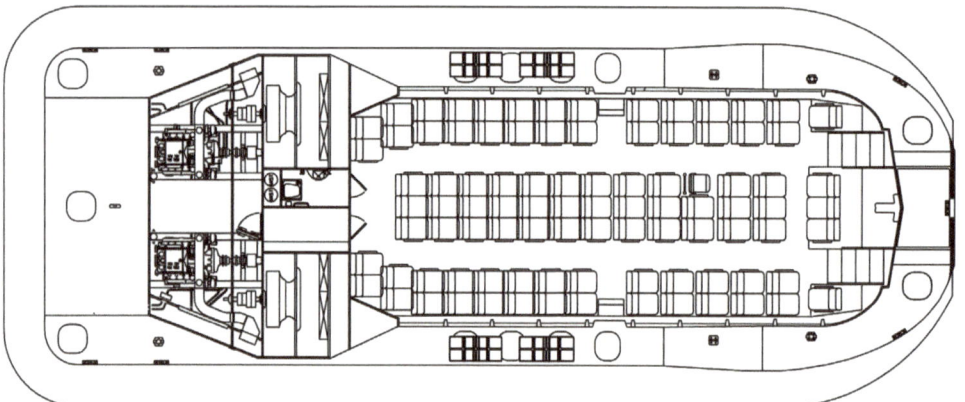
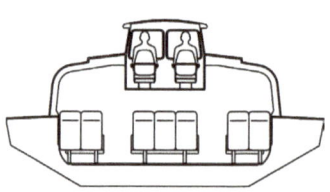

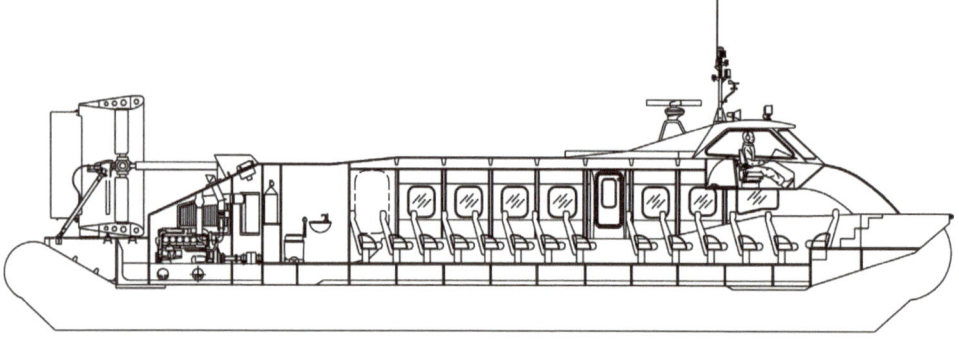

(Above) The ABS P-92 is a modern means of transport that complies with current licensing regulations. Its shielded silenced diesel engines, plastic construction and shrouded propellers make it no noisier than a modern truck.

Unlike most other hovercrafts of the same size, it offers a toilet and a sound-proofed engine room that can be entered during the ride.

Technical data:

Length:	21.61 m
Width:	8.83 m
(dimensions incl. aprons)	
Hovering height: 1	m
Passengers:	up to 92
Propulsion and fan:	2 diesel engines of 791 hp each
Speed:	40 kn

6. Riding on an air bubble

The completely amphibious hovercrafts are certainly spectacular and fascinating, but they are in the minority in numbers compared to their „half-brothers", the ships supported by air cushions.

With this technology, there are two hulls, usually in the shape of a catamaran. Air is pumped between them by a fan to reduce their draught to the minimum possible.

To prevent the air from simply disappearing, there is a flexible curtain between the hulls on the foreship and on the stern (fore and aft), which is also called an „apron" here. The two hulls usually dip more than a metre when stationary, but when the blower is switched on, the draught is reduced to just a few decimetres. The water resistance in the „inflated" state is then so low that a high speed can be achieved with relatively low power requirements.

Due to strange coincidences in the development of a terminology for the new ship categories, it was decided to call this category „SES" or in long form „Surface Effect Ship", a term that basically means nothing. Nevertheless, we will continue to use it here so as not to violate naturalised standards.

These SES appeared on the market in larger numbers from the mid-1960s onwards. But they were mostly small passenger vehicles with no more than 70 seats. They were less sensitive in operation than hydrofoils and gained a market niche mainly in water bus transport in harbour areas and other sheltered waters.

They required less energy input for air supply than hovercrafts and could be moved and steered by propulsion systems acting in the water, such as propeller systems. This protected the environment by emitting very little noise, which had always been the main problem with hovercrafts. Later, mainly Norwegian manufacturers began to offer larger SES for open waters. With speeds of up to 45 knots in some cases, these were often extraordinarily fast and soon began to become serious competition for the hydrofoils that had dominated the open seas until then.

The Norwegian company Ulstein AS, now part of the Rolls-Royce group, offered a whole series of designs of varying sizes. Brodene AS and Vestamarin AS were also Norwegian shipbuilders who excelled in this field.

SESs could sail comfortably and quickly in shallow and sheltered waters. But as soon as an SES headed for the open sea with its higher waves, its good qualities could quickly disappear. One of the worst effects was the so-called „Copplestone Effect". But what is this?

You have to imagine that an SES hardly displaces any water during high-speed travel, but rather glides over it riding on an air bubble. Only the very thin catamaran keels on the sides protrude below the water surface.

During the journey, waves shorter than the length of the boat get into the hollow space between the keels, because the front skirt does not exert enough pressure on the water surface to „level" these waves.

These small waves cause the air in the cavity to suddenly compress at high speed and as they occupy part of the air volume.

This compression of air is passed on to the hull of the ship, which receives a blow as a result. The rapid succession of subsequent waves ensures that these blows build up to a drumming that actually bears a resemblance to the behaviour of a car driving over a rough pavement. Neither people, material nor machines can withstand this jarring over a longer period of time, and that is reason enough to reduce speed.

Now a fast ferry travelling slowly is a contradiction in terms. Therefore, a great deal of research was done from various sides to eliminate this effect. The solution finally

(Above) A view underneath an SES illustrates how it works. The side walls of the fuselage generate part of the lift and seal the air cushion to the sides. A flexible skirt at the bow and at the stern ensure that no air can escape from the passenger compartment.

consisted of a set of valves that discharged the pressure surges to the outside air under computer control. Thus, whenever there was an abrupt increase in pressure, the valves were opened. Since waves do not occur regularly, a considerable technical effort was necessary to adjust the valve opening phases to the occurrence of the pressure surges. The SES also lost buoyancy air through the valves, which had to be pumped back in by a higher blower output.

SESs also have to be built very light to minimise the propulsive power needed to maintain the air cushions. Unfortunately, lightly built sea-going vessels are usually not very robust either. In addition, the front and rear skirts of the SES wore out regularly and had to be repaired at some cost. When efficient conventional catamarans appeared on the market in the 1990s, the SES were gradually replaced.

Today, hardly any SES ships are built. The most spectacular example is the „Techno-Super-Liner" from Japan, which was launched in 2004. With a length of 140 metres and a speed of almost 50 knots, it is one of the largest and most powerful fast ferries in the world. Its sheer size compensates for the vibration problems known from older SESs in rough seas.

The SES is predicted by some experts to have a great future in the development of fast ships beyond 100 metres in length. But there are alternative successful technologies that are also seen as a panacea for creating oversized fast ships.

Increasing buoyancy through air cushions has not yet been shelved; rather, new and surprisingly simple solutions have emerged. They are able to compensate for the main disadvantages of the SES.

We are talking about hulls of catamarans and also mono-hulls supported by air cushions, which are being worked on in Russia, the Netherlands and the USA. They could have the potential to raise the current general economic speed limit of fast ferries from 40 to 50 to 70 knots.

Imagine a catamaran hull where the underwater part of the hull has been cut off at the waterline. Air is now blown into this cavity. The ship can then float and sail again, and

the frictional resistance of the hull when cutting through the water surface is almost eliminated - the vehicle glides on a cushion of air. Since this is narrow and thin, no wave can go underneath and thus cause the „Copplestone Effect".

If you omit cutting the fuselage at the bow a little, you can connect a wave-splitting front fuselage with the air cushion part. This makes for a craft that generates little drag, does not exhibit particularly negative behaviour in high seas and, moreover, as an SES, does not require a high-maintenance skirt or a lot of air. This concept is being pursued by some designers in the USA and Europe.

but so far only one 27-metre catamaran, named „Purr-seaverrance", has been built.

The sidewall hovercraft concept is far from extinct. In Norway, a new company called ‚SES-X Marine Technologies' has set out to enrich the world of fast ferries with electrically powered speedboats that ride on a cushion of air.

Instead of a smooth ship's bottom, the single-hull vehicles have a hollow space open to the rear into which air is blown by blowers. This reduces the boat's friction with the water considerably, without the need for a floating effect. The modern electric drive makes it possible to speed through the water at about 30 knots without a guilty (environmental) conscience.

In Norway, fast ferries are an important part of the infrastructure, just like buses on land. Instead of having to drive around an often very long fjord on country roads, it is simply crossed quickly. But the diesel-powered catamarans and monohulls of the past have to be replaced by emission-free solutions in the new era. In a country that promotes electric-powered transport more than other countries around the world, propulsion by battery power is a natural development. Meanwhile, SES-X also offers workboats that could be considered for offshore wind farms near the coast, for example.

The design of SES-X is in line with the ideas of Roger Thornycroft, the pioneer of fast ships, from the late 19th century. He would certainly have welcomed the idea of electric propulsion, because at the time, the engineering

world was not as mentally fixated on the internal combustion engine as it is in the 20th century.

Amphibious hovercrafts, on the other hand, always had a hard time. The technology never really proved its high potential, which was initially acknowledged in the 1950s. The crux of the matter is that they emerged at a time when people were learning that crude oil-based fuels were a limited resource. Only users for whom operating costs are not a major concern, such as the military, can afford to use large hovercrafts today.

In the civilian world, use is essentially limited to small and medium-sized amphibious vehicles with lengths of up to 22 metres. Really large hovercrafts are hardly economically viable today. However, a new foray into size has been made with the new BHT-130 from the Isle Of Wight. It was only possible due to the very powerful light diesel engines available today. But this innovative vehicle has also been stored since 2011. The traditional hovercraft ferry route between Portsmouth (UK) and the Isle of Wight

is now served by much smaller vehicles.

The number of active hovercraft in civilian use is marginal. . Fewer than ten are currently used internationally as fast ferries. Hovercrafts are much more successful as work and rescue vehicles.

But one should not complain too loudly: compared to the number of commercially used supersonic aircraft, this is still a very good result.

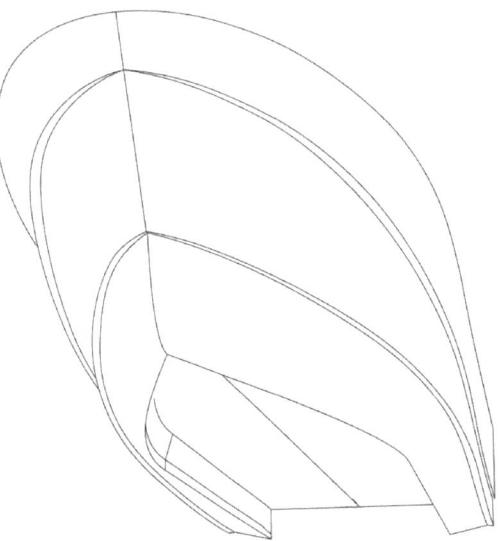

(Top) A current design of SES-X: A monohull fast ferry for up to 74 passengers with a length of 20 m, width of 6.1 m and a speed of 25 knots - powered by battery electricity.

(Beneath) A look under the SES-X: clearly visible is the cavity that holds the air bubble. This concept is also called „air lubrication").

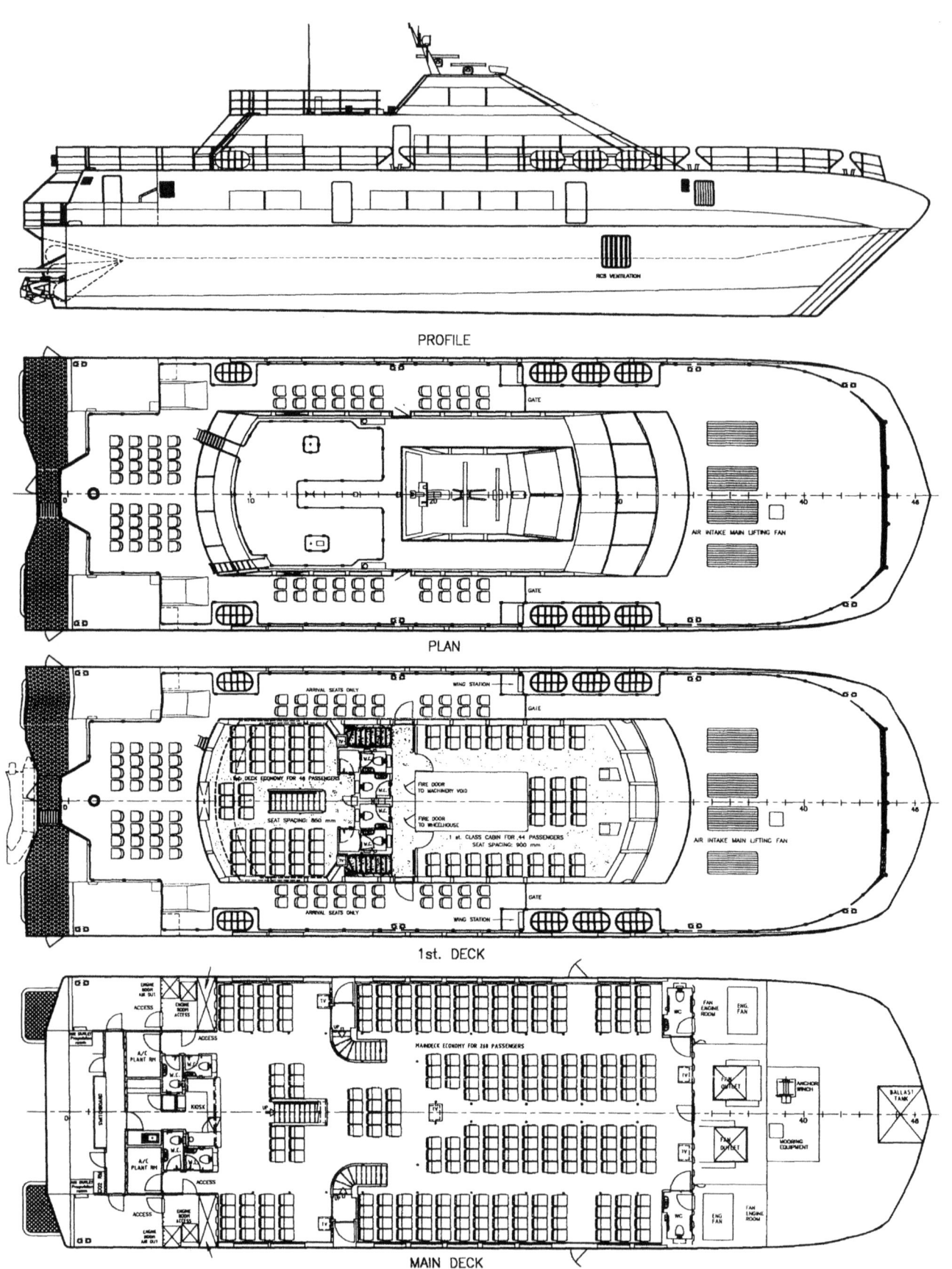

PROFILE

PLAN

1st. DECK

MAIN DECK

(Above) The ULSTEIN UT-928 is a typical SES design from the 1980s. It shows all the characteristic assemblies of an SES. Many SES were built from fibre composites, as any weight saving was badly needed. This is the „Tamahine Moorea" built under licence in Australia.

7. On one, two or three hulls

It is now appropriate to look at the various designs that have developed in the fast ferry industry.

The hull shape is decisive for assessing a design in terms of its usability in a particular sea area and its economic efficiency. It determines what propulsion, what power is needed, what downtime due to bad weather one has to accept, and how expensive the new construction will be.

Hardly any other trend in shipping technology has generated such a large number of innovations and inventions as that of having to go fast. There are literally several hundred concepts for shaping a hull, and many new ones are added every year.

But for all the inventors' ingenuity: the number of shapes that are used in practice is quite small. That is why only the most important ones are presented here. The English word „monohull" literally means „one-hull" and that is what it means. Every normal ship is basically a monohull. With fast ferries, however, there are still some special features to consider. When a ship increases its speed more and more, a bow wave develops that piles up higher and higher. Its resistance adds to the already existing frictional resistance of the hull. At some point, the ship reaches a speed above which it cannot accelerate any further without having to overcome the resistance of the bow wave or somehow avoid it. This speed is called hull speed.

If a small boat now tries to break through the bow wave by using even more propulsion energy, it may become unstable or suffer damage in some other way. A large boat usually begins to vibrate, but does not suffer an accident. But any further application of energy will not bring faster progress, for the hull must fundamentally change its condition as a body displacing the water. One possibility is that it transforms itself into a so-called glider.

In this case, the bow climbs over its own bow wave and the hull glides on the water from now on instead of penetrating it. The water masses now flow along under the hull of the ship and carry it by the pressure that their current exerts on the underside of the hull. This buoyancy is what is known as „dynamic" buoyancy. Of course, this only works if the mass of the ship is less than the pressure of the water it is gliding on. It is obvious that a large tanker, for example, could never glide even with the greatest effort. Here the mass of the ship would always win against the dynamic water pressure. But it is a different matter with a light motorboat. Therefore, the hull of a fast monohull is preferably equipped with a flat bottom and must also be light in construction.

(Above) Two 44m catamarans from AUSTAL Ships meeting in spectacular way.

Ideally, the monohull is quite wide so that the gliding surface is as large as possible. Otherwise, dancing on its own bow wave would be a pure balancing act. There are various types of planing hulls, such as the half-planing hull, which only planes with the foreship, but is a displacement hull aft. Similarly, there are V hulls, which have a V cross-section instead of a flat bottom. However, the functional principle is always the same.

A monohull must be completely inherently stable as a ship. It cannot rely on supporting auxiliary hulls. This means that the ship is naturally wider, which in turn creates quite a lot of resistance. In addition, a rather large „lump" must be driven through or over the waves. The monohull cannot simply cut through the waves as many catamarans can. Instead, it has to glide on the water.

Therefore, more power is needed to propel a monohull at rest than a comparably sized catamaran.

However, monohulls are technically easier to build and handle. The designers can fall back on time-honoured know-how, which also facilitates the technical approval of these hulls. Since all propulsion units are installed in the same hull, auxiliary units such as pumps etc. do not have to be installed twice in the same vessel. This is different from a catamaran, which has its own engine room in each hull.

It is noticeable, however, that monohulls are mostly only used where sea conditions and waves are less dangerous. This is due to the fact that monohulls tend to

This is because monohulls tend to roll and often have little ability to cut through waves. The catamaran, on the other hand, is an ancient type of ship that was virtually „rediscovered" in the later 20th century.

Its principle is based on distributing the mass of the ship over two hulls lying next to each other. Therefore, it is unimportant how inherently stable one of these hulls is. So both can be built very slim and have correspondingly less flow resistance. Likewise, these two very slender hulls are much better able to push through their own bow wave than the one hull of the monohull. Up to a certain speed, which is consistently higher than that of a monohull of the same mass, the catamaran does not need to mutate into a glider. This has the advantage that it does not have to jump over waves, but rather cuts through them. Anyone who has ever been on a motorboat moving very fast through a choppy sea knows that this jumping over the waves can lead to very unpleasant knocks. On a catamaran of the same size, the ride would be smoother.

Most fast ferry catamarans are still displacing water at high speed or at best only partially planing. Therefore, they can still sail safely in much higher waves. Some designs of catamarans are even designed to barely rear up when they come into contact with a high wave, but to pierce it like a spear - the so-called „wave piercers".

Wave piercers differ from normal catamarans only in the shape of the foreship. The tips of each foreship of a hull are pulled strongly forward in the area of the water-

(Above) A view under the platform of a modern passenger catamaran. The „false bow" on the centreline breaks up waves that would otherwise hit the underside. It also increases buoyancy when the ship threatens to bore into a wave.

line. This pierces the wave. Between the two hulls is the so-called „false bow". This normally has no contact with the water, unless the wave is a little too high. Then the false bow splits this wave and pushes it into the catamaran tunnel. The wave can now no longer develop its effect, it has been „flattened", so to speak.

If everything works correctly, the wave piercer can prevent the foreship from rearing up and thus the unpleasant pitching by pushing through and splitting the waves. In practice, it doesn't work completely without help. Even the largest wave piercers often need the support of stabilisation systems with computer control located at the bow.

It is noticeable that wave piercers are most often used in sea areas that frequently experience bad weather. Until now, this design has only been used on very large fast ferries, because it was easier to adjust the hull shape precisely to the wave profiles to be passed through.

But recently, suppliers of smaller ferries and motor yachts have begun to take advantage of this effect as well. Unlike other more complex designs, wave piercers are not particularly expensive to build and are just as economical to operate as catamarans without this bow shape.

One particular type of catamaran design forgoes many of the advantages that can be gained from slender hulls

and the glider principle. These are the so-called „semi-submersibles".

While normal catamarans often have only a shallow draft, semi-submersibles are virtually dependent on a large draft. They gain their stability from that static buoyancy which is based on the balance between the mass of a submerged body and the mass of the water it displaces. This principle is the same that allows a submarine, which is precisely leveled to a certain water depth, to float stably. If the submarine becomes heavier, it sinks; if it becomes lighter, it rises.

The typical semi-submersible has two deeply submerged cylindrical hulls, which are connected to a platform above the water by supports or side walls that are as thin as possible.

The submerged depth is adjusted to a certain value depending on the mass of the ship. This remains constant as long as the mass of the ship does not change, for example due to fuel consumption.

It is particularly advantageous for the semi-submersible if it has a greater draught. Then the part of the ship that generates the most buoyancy, the cylinder hulls, are so deep below the surface that the swell can no longer have any significant effect on the ship.

Only the thin supports or side walls, which hardly contribute to the buoyancy of the ship, are in the area of effect of the swell. For a semi-submersible, which is about 30 metres long, a draught of 2.5 to 3 metres is not unusual. The torpedo-shaped tubes are already about 1.5 to 2 metres below the effective range of a swell of 1.5 to 2 metres average wave height. The catamaran is therefore considerably less affected by the swell than a conventional hull of the same length, which has a draught of only 1.5 metres. At greater wave heights, the semi-submersible begins to move with the swell, but this happens in a much more subdued manner than with a conventional ship. This unusual type of ship is therefore the right address for all notoriously seasick people!

The concept described here has been commonly referred to as SWATH - i.e. Small Waterplane Area Twin Hull. So far, only a few SWATH fast ferries have been built. The

reason is that this hull form requires very complicated propulsion systems and is also particularly expensive to assemble. Either the drives are housed in the floats, where there is little space available, or long shaft systems or complex power transmissions are accepted to transfer the power of the engine to the propeller on each side. SWATH are also vulnerable in other respects. They do not use the so-called „form lift" to stabilise the horizontal position, but only the static lift, which can have considerably less effect as a force when the position changes.

Shape buoyancy can be understood in such a way that, for example, a ship with a V-shaped cross-section generates decidedly more buoyancy when it dives deeper. This

(Below) On this Wave Piercer model, a fold-out T-foil has been placed under the centre bow to dampen vertical movement when riding through high waves. The comfort of the passengers and the seaworthiness increase considerably as a result.

is not the case with a SWATH, because its cross-section is approximately the same in the front, middle and back.

in the middle and at the back. This means that the balance has to be kept upright all the time, even when the boat is loaded with ballast. This makes the SWATH shape particularly suitable for ships that sail without a large load, for example as pilot boats, research vessels or for the military. Another disadvantage is the very high water resistance of this hull form, which is more of this hull form, which requires more power than a normal fast ferry of the same size.

Most SWATH ships are not faster than 22 knots - many are even much slower. The original form of the SWATH did not yet satisfy the ferry builders very much. But they

also wanted to keep the positive aspects, like the good behaviour in bad weather. So a compromise was devised between the catamaran hull and the SWATH.

which today is generally referred to as a „semi-SWATH". In it, the props have been substantially modelled on a ship's hull in terms of their shape, which is why they can now also develop form buoyancy. Nevertheless, the hull still has quite a large draught, so that advantages are also gained by the SWATH principle. Depending on the draft and the transitions between the submerged part of the hull and the upper part, the result is a catamaran that can sail calmly through the sea, but does not have as much resistance or is as sensitive to shifts in the centre of gravity as the purely as the pure-blooded SWATH.

This compromise proved to be extraordinarily successful. For example, the Australian company AUSTAL Ships built a large number of fast ferries as semi-SWATHs. The latest entry in the semi-submersible race is SLICE, and was developed by Lockheed-Martin in the USA. Instead of two cylindrical floats, four teardrop-shaped floats were mounted on each platform. The propellers were also at-tached to the two in front. This strange vehicle proved to be extraordinarily successful. The seaworthiness was in some cases even better than that of a SWATH and the speed that could be achieved with a still economical use of power was considerably higher at about 30 knots. Lockheed-Martin therefore received an order in 2004 to build six supply boats, each 26 metres long, for the Mexican PEMEX oil company. They reach a cruising speed of about 24 knots. The first of the series was successfully tested at the end of 2006. SLICE is currently considered the highest development stage of the semi-submersible hull. Overall, the SWATH concept has not proven itself. Its apparent seaworthiness fades away when wave heights reach the lower edge of the platform. Almost all commercial SWATH concepts have failed and never found their way into large-scale production. On the other hand, hybrids like the „semi-SWATH" have been quite successful.

No one forbids building ships with even more than just two hulls. The next candidate on the list is therefore the trimaran with three hulls. It has a large central hull and two lateral support hulls.

(Below) A typical „wave piercer": The aggressive-looking „false" bow is able to break up high waves and push them into the tunnel between the hulls. Smaller waves are simply pierced by the bow tips of the catamaran hulls - in other words, „pierced".

Due to the safety provided against capsizing by the two side hulls, the centre hull can be made very long and particularly slender, because it does not need any inherent stability. It thus achieves more ideal resistance values than a monohull and can thus reach a high speed while saving fuel. The long, narrow hull is ideal for cutting through waves, which is why the trimaran is also very seaworthy. The two side hulls may have significantly smaller dimensions than the main hull. Their sensitivity to waves is thus also low, because they can also be built very long and slender.

Thus, the trimaran is actually an ideal design for fast ships, but strangely enough, very few shipyards noticed this until now.

(Below and right) The SWATH „Cloud X" is one of the few fast SWATH ferries ever built. Powered by two Avco-Lycoming gas turbines, she reaches about 30 knots. Due to the SWATH hull shape, her passengers do not get seasick even in wave heights of 2.5 metres. She was planned as a casino ship. Today she can no longer be found but is probably mouldering away in some harbour. The fuel consumption of the two gas turbines was extremely high and the ship quickly became uneconomical. The ship quickly sank into oblivion and was scrapped a long time ago.

Length:	*35.5 m*
Width:	*18.30 m*
Draught:	*2,3 m*
Propulsion:	*2 X 4,000 hp*
Cruising speed:	*30 kn*

The first SWATH fast ferry was the MESA-80, built by the Japanese Mitsui corporation in 1979. It was used in rough Pacific waters on lines in the Japanese island world. It could reach up to 28 knots and still sail comfortably in wave heights of up to 4 metres.

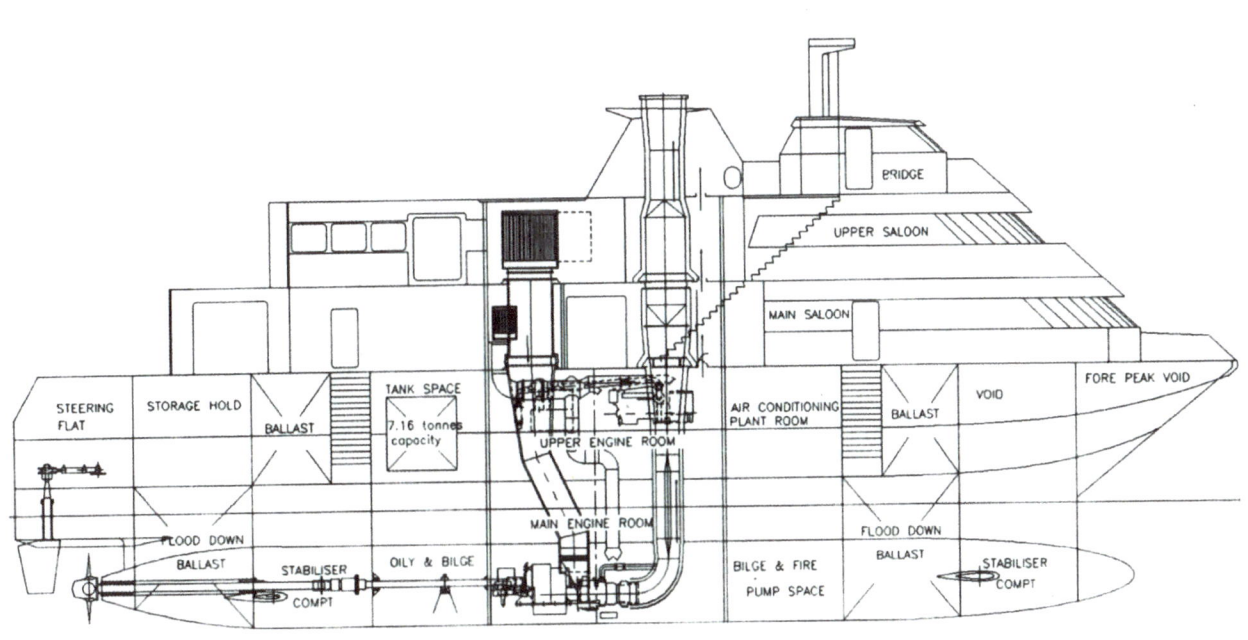

69

(Above) This is a so-called „FOB SWATH" - the last nomination in the SWATH world. He has a special characteristic, because he can change his draught. When it surfaces, as seen here, it is not a SWATH, but just a normal fast catamaran. When it dives deeper, it can use the SWATH properties and provide a stable platform when it is moored to a wind turbine. In this way, water resistance is reduced during the crossing and operating costs are lowered. There are now already five boats of this type built in Northern Denmark since 2012.

It seems that almost everyone in the industry was totally fixated on catamarans. Once again, Australian shipyards stepped up to solve this problem. With the „Triumphant", a 58-metre long catamaran was launched around 2003, which proved to be a ship for bad weather right from the start during its test runs. The builders were also able to prove that this ship required around 30 percent less fuel than a catamaran with the same payload. The „Triumphant" needs three 16-cylinder engines with 2000 kW each, whereas the comparative catamaran needs four engines of this type to reach the same speed!

But that was not all the industry was capable of: Lineas Fred Olsen ordered a replacement for the last large conventional ferry in its fleet of Canary Island traffic around 2002. Fred Olsen knew the value of large fast ferries very well, as his company had already been successfully using several large catamarans since 1998. Now a fast ferry with an even larger payload was required, which should also be better able to cope with the often high Atlantic swell.

The elongated hull of a trimaran seemed to be the best solution. The high slenderness of the centre hull and the great length of 127 metres that was planned would be able to breeze effortlessly through the waves of the Atlantic and still carry a large payload. AUSTAL Ships' design department approached the matter systematically: They first built a scaled-down model that offered space for two people and was propelled by an outboard motor. This vehicle could now make its rounds in a bay near Perth, Western Australia, with all possible parameters being measured. The design department was now able to design a gigantic trimaran that could achieve a very high transport performance of up to 40 knots.

Although the state of the art in fast ferry technology has reached a respectable level by now, there are still technologists who see further need for improvement. One such was at the University of Stellenbosch in South Africa. There, Professor Stefan Hoppe developed the technology that was to prove to be another revolution in fast ferry navigation in experiments that lasted for years.

Hoppe first combined small hydrofoils with glider hulls from motorboats. These wings lifted the foreship of these boats a little out of the water, which reduced drag. It worked best with catamarans, because the mounting of the wing as a transverse connection between the hulls was ideal. The maximum speed was increased by up to 25 per cent or the energy consumption was reduced by up to 25 per cent while the speed remained the same.

At the same time, the seaworthiness of the boats increased considerably: because the dynamic lift generated by the wings when diving into the water is many times more effective and also comes into play more quickly than the static lift generated by the hulls, the wings are quicker to soften the impact of diving into a wave. The new concept was named „HYSUCAT" for „HYdrofoil SUpported CATamaran". Since small catamarans have long been built as sports and fishing boats at the Cape of Good Hope, it quickly found favour. Nevertheless, it took more than 15 years to spread HYSUCAT worldwide.

Like all good inventions, it is now being diligently copied. Due to technical and economic limitations, the hydrofoil support for catamarans can only be used for ships with lengths of up to about 40 metres, because on even larger craft the hydrofoil would be too large and would create too much resistance of its own.

The HYSUCAT principle has now been adopted or further developed by many suppliers worldwide. Even in trimarans such as the one on the already

Triumphant", it has been used to good effect. Here it was able to improve the already good characteristics of the trimaran even further than was possible with catamarans. In the meantime, the auxiliary wings are sometimes fitted with steering fins to suppress the pitching and rolling of the craft.

Every year, new ideas and concepts spring up like mushrooms to improve seaworthiness, economy and performance in maritime shipping. But the competition is merciless and only a very small percentage of these ideas survive their first year. In many cases, the concepts range from only „conditionally usable" to the category of „total humbug".

Ultimately, the introduction of a new enabling technology requires a lot of convincing and practical proof of feasibility to convince investors. In contrast, it is often inexplicable to the proponents of new methods why the

investors do not enthusiastically support the positive features of their concepts, but instead adopt a wait-and-see and cautious attitude. This dichotomy between technologists and financiers is often not so easy to bridge. But technology enthusiasts should bear in mind that promises, paper proofs or even laboratory tests are not enough to create confidence in a new technology in which two or three-digit million sums are to be invested. The risks in the fast-track business are too great to allow light-hearted chasing of new trends.

(Above) A "The-Artist-Impression" of the SLICE crewboats built for the Mexican PEMEX Group.

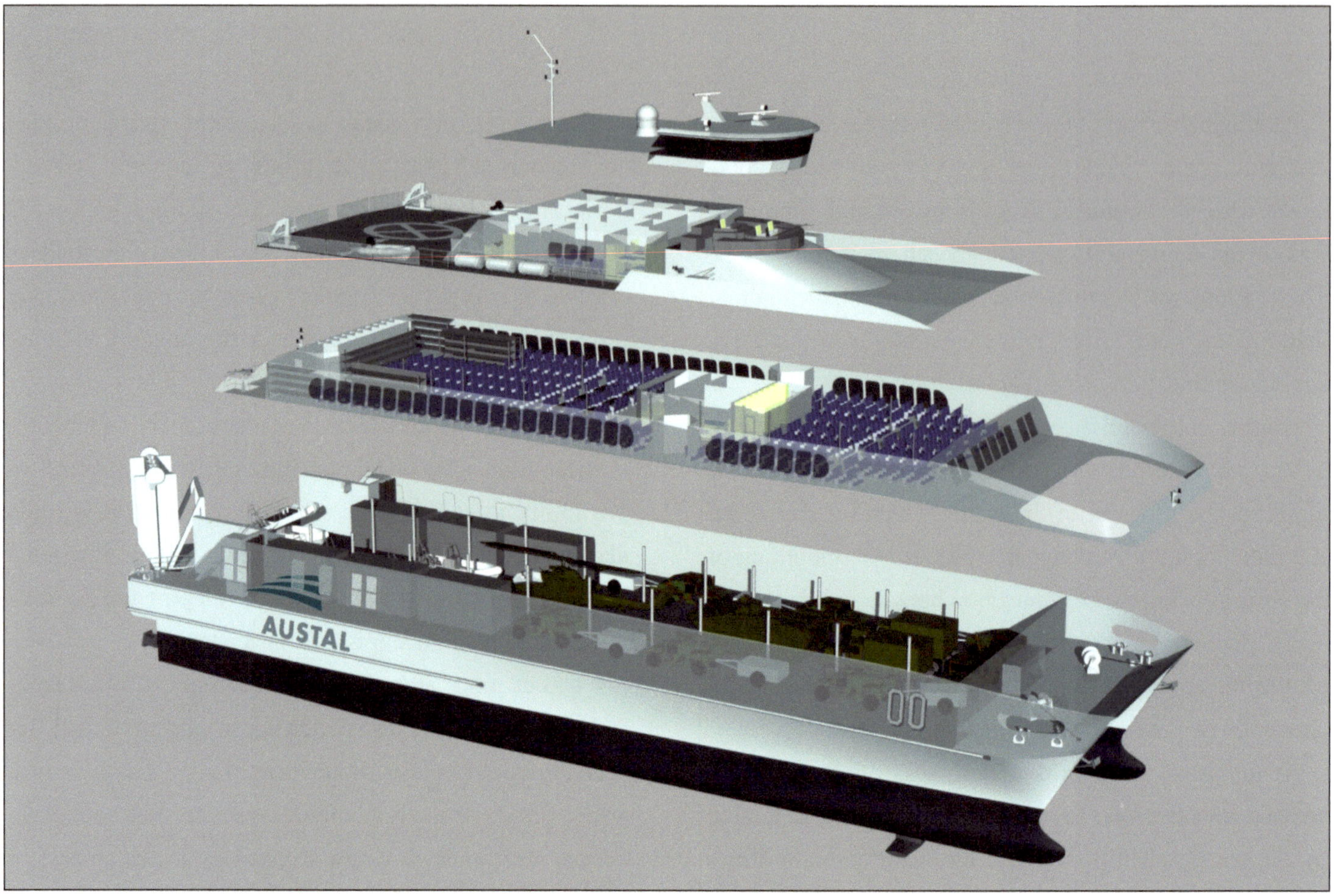

(Above) This 72m catamaran of the new „Spearhead" class has been developed by AUSTAL SHIPS for the transport of troops and their equipment. The vessel's large draught, which is particularly noticeable on the forecastle due to the bulbous bow, marks the design as a semi-SWATH. Due to the bulge bow and its draught, this catamaran is designed to push through the waves rather than glide over them. The high speed is achieved solely by the great slenderness of the hulls.

In contrast to the pure SWATH, the transitions between the underwater hull and the part above the waterline run in smooth curves. This means that the catamaran also has the form buoyancy to stabilise itself in high waves. Computer-controlled stabilising fins under the foreships support it in this. They are often called „T-fins" because of their shape. AUSTAL supplies a number of larger transport catamarans to the US Navy.

(Left) One of the first fast ferry trimarans built worldwide is the „Triumphant", built by NWBS-Ships from Tasmania. Not only the hull itself is a technical innovation, but also the novel support of the hull lift by hydrofoils. The manufacturer was able to prove that the same number of passengers can be transported with the „Triumphant" in a way that is about 25 percent more fuel-efficient than with a catamaran of the same size and speed.

Length:	54.5 m
Width:	15.2 m
Draught:	2.1 m
Propulsion:	3 x MTU 16V 4000 M70
Total power:	6,960 kW
Passengers:	463
Speed:	40 kn

(Right) David against Goliath?
AUSTAL's 127m trimaran turns in front of an astonished motorist.

(Below) The vehicle deck of the 127-metre trimaran open to the rear as it manoeuvres backwards towards the ferry ramp.

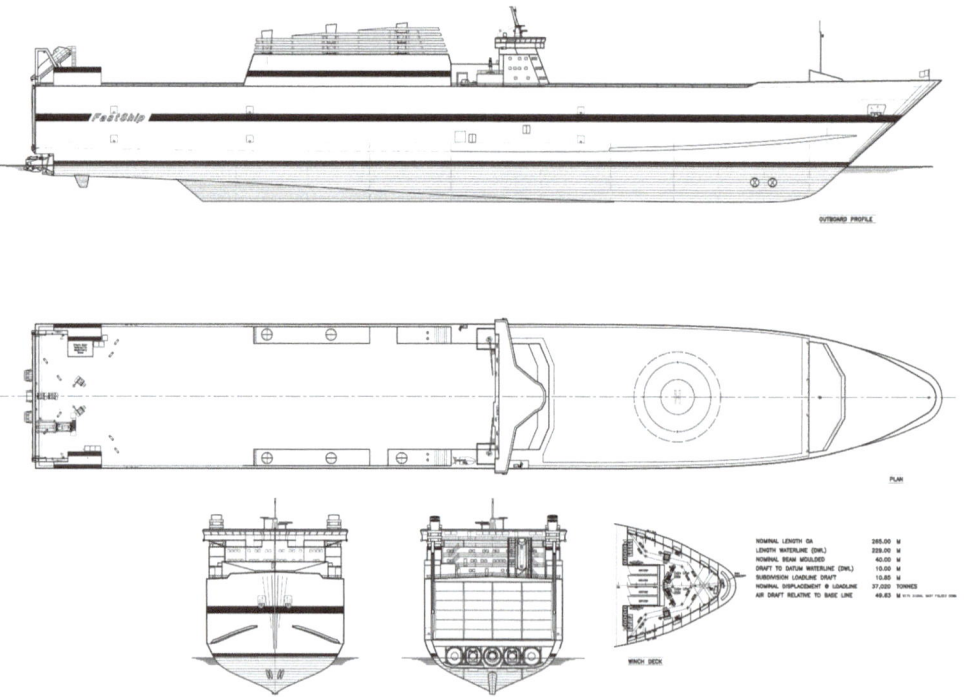

The catamaran „Spirit Of Ontario" during the high-speed test. The catamaran was originally delivered in April 2004 for the Rochester (New York State) to Toronto route on Lake Ontario. Extreme problems with the authorities, the unions, and other calamities prevented a timely commissioning and reliable operation. After the bankruptcy of the operator in April 2007, the vehicle was sold to the Flensburg Seetouristik GmbH shipping company for use between Spain and Morocco. The ferry is one of the fastest large catamarans ever built. Today, after further changes of ownership, she sails under the name „Virgen de Coromoto" on a route in Venezuela. Whether this is still the case in view of the situation there could not be determined. Her fate is typical for the world of fast ferries. Due to the high acquisition and operating costs, many ambitious projects often fail in the first five years due to lack of liquidity. Key Data: Length 86.6 m; width 23.8 m; passengers 774 ; cars 238; propulsion power 32,800 kW; speed up to 45 knots.

The "Fastship" would have been the largest civil fast vessel ever. It was planned for fast atlantic crossing between the old and the new world to compete against the air cargo carriers on this route. But the 262 m long vessel was never built.

It was planned to install 5 Trent gasturbines with total 250 MW. KaMeWa designed for Fastship the largest waterjet ever with a diameter of 3.25 m. The cruising speed would have been 37 Kts.

The program costs estimate was about 200 mio. USD.

(Above) The catamaran HMAS „Jervise Bay" was in service with the Australian Navy for some time as a superfast transporter. It triggered a kind of boom for this type of ship in the military.

This is a so-called INCAT-86 (length 86 metres), which was originally developed as a civilian ferry.

(Below) The „Littorial Combat Ship" (LCS) USS „Independence is type ship of a new class of fast combat ships of the US Navy. Length: 127.2 m

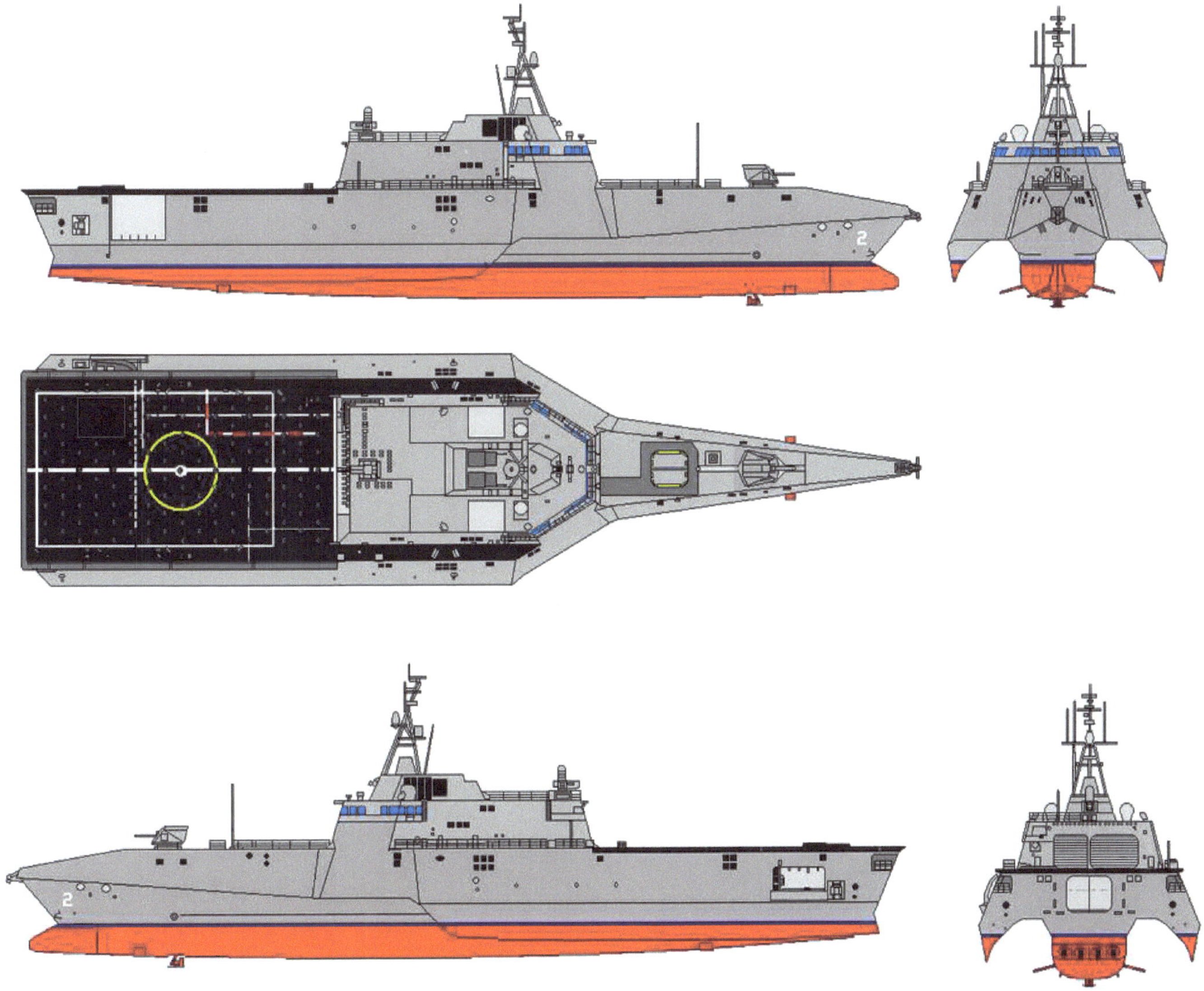

The Hybrid - a typical HYSUCAT

The small catamaran „Nordstern" is one of the first fast ferries built in Germany. During its construction in the early 1980s, the latest findings were immediately used to achieve an optimal result. Using the then new hydrofoil support, a compact vehicle was developed and built in Bremerhaven, which despite its low propulsion power of two times 662 kW (887 hp) could still accelerate to 36 knots cruising speed. The ship was to sail from Bremerhaven to various East Frisian islands. The hull was designed as an asymmetrical catamaran, i.e. the inner side walls of the hulls ran almost vertically upwards, while the outer ship's walls were adapted to the shape of the boat's hull.

Between the keels, two wings made of high-quality steel were mounted to support the hull's buoyancy. Another new feature was to install the wheelhouse at the stern. This created a passenger space not interrupted by stairways and supports. Despite some incidents of a financial nature, the „Nordstern" was completed and tested in August 1994. During one of these trips, the shipyard had a powerful motor yacht accompany the ship so that photos of the new ship could be taken from there. But the „Nordstern" simply outran this yacht!

The HYSUCAT principle proved itself. The seakeeping behaviour of the rather small catamaran was good, because the hydrofoils were not only advantageous in terms of economy, but also had the effect of dampening the pitching behaviour. The Nordstern was used in the seaside resort service for several years from 1995. Unfortunately, the operating costs were too high to be covered by the revenues. In addition, despite the HYSUCAT principle, the catamaran, which was just over 20 metres long, was simply too small for the sea state of the North Sea. There were too many cancellations of crossings due to bad weather.

So the ship was finally sold abroad in the late 1990s. She is still sailing somewhere and turns up on the second-hand market from time to time.

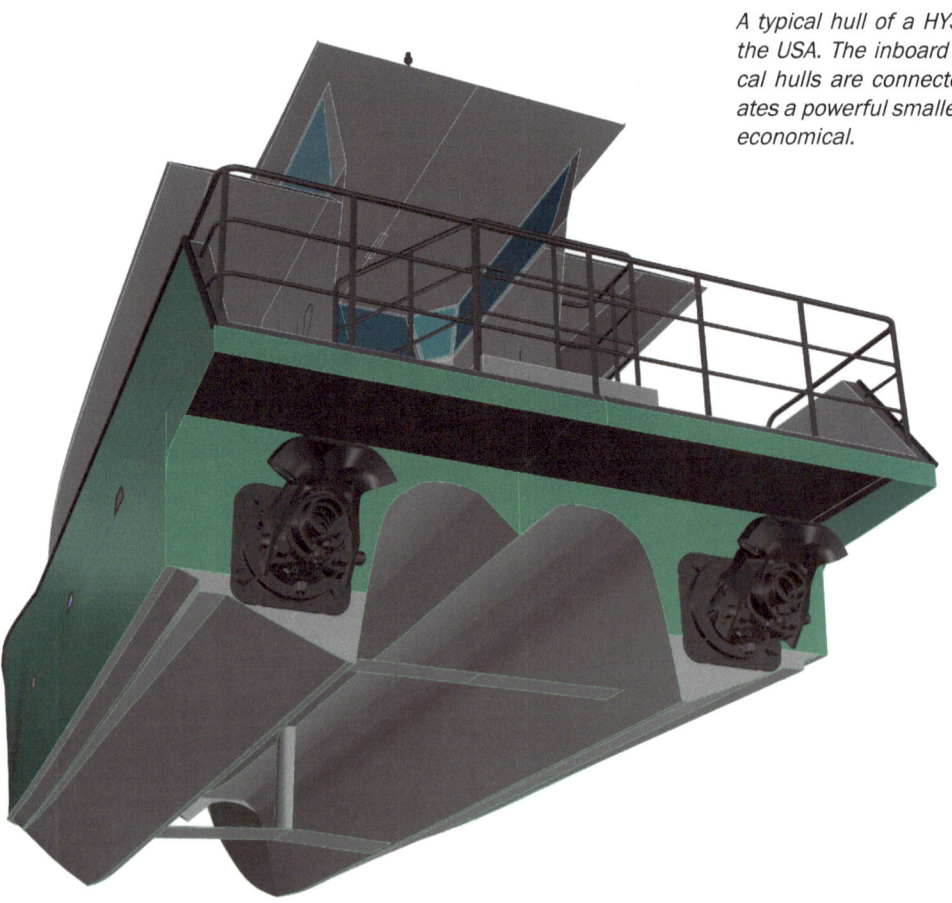

A typical hull of a HYSUCAT boat designed in the USA. The inboard keels of the asymmetrical hulls are connected by the fins. This creates a powerful smaller boat that is robust and economical.

In the meantime, the principle of the HYSUCAT is in use all over the world, as it is a cost-effective method of improving the sea-keeping qualities and speed of a catamaran. No mechanics or electronics are required, just well-made stainless steel.

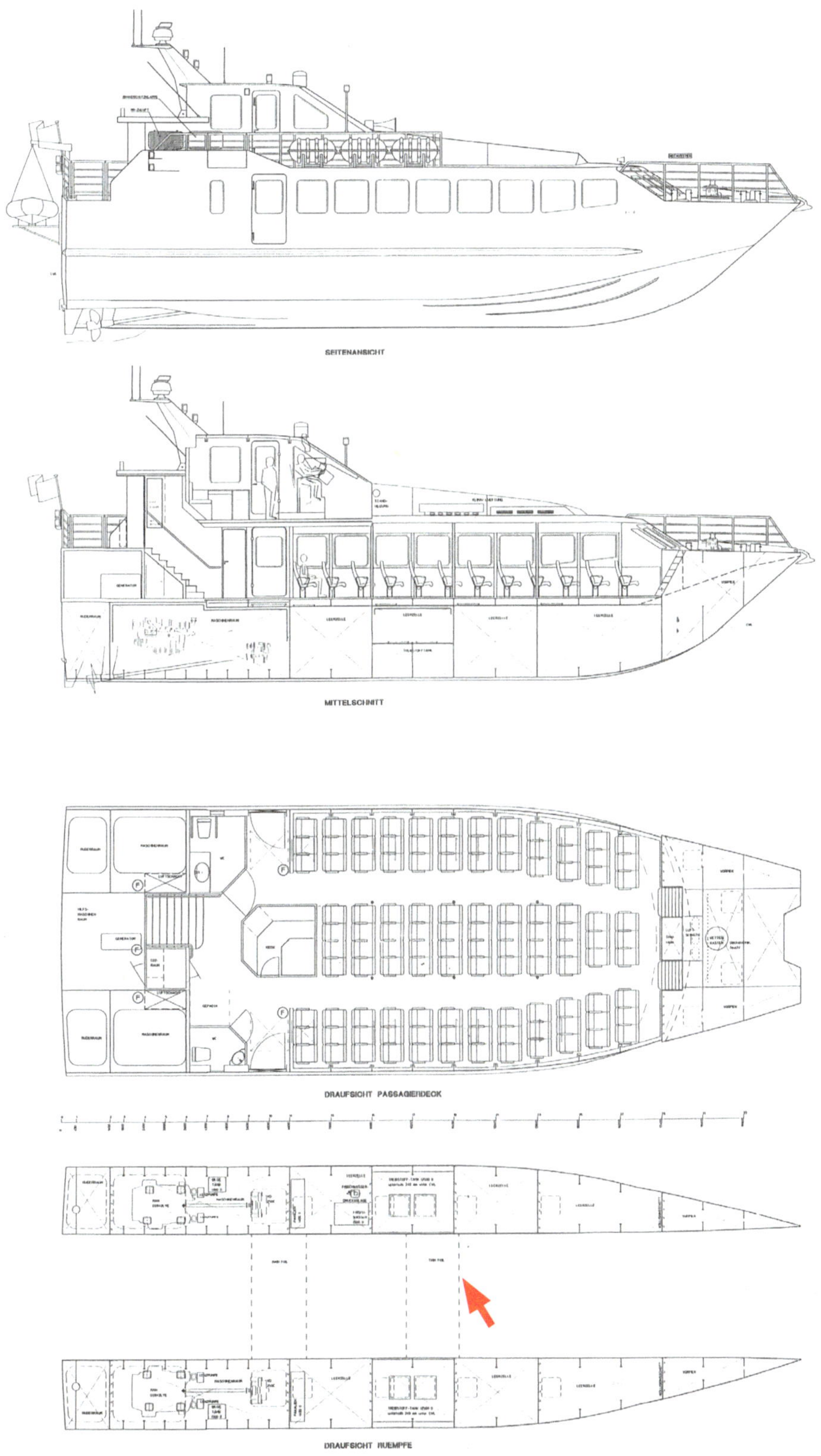

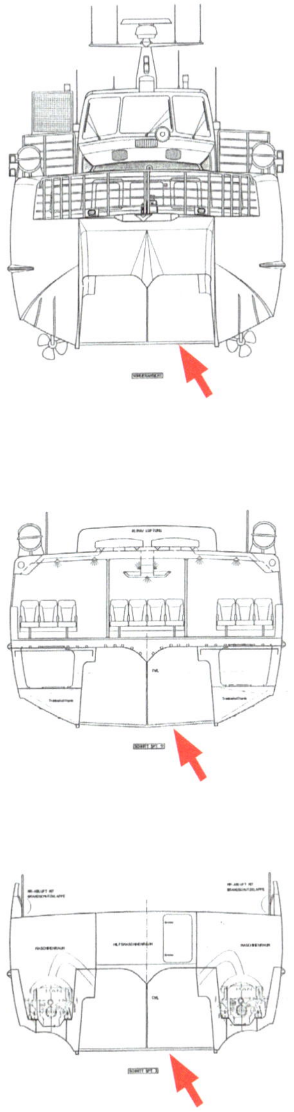

The „Nordstern" in facts and figures:

Length:	24 m
Width:	7.35 m
Draught:	1.43 m
Displacement:	55 to
Propulsion:	2 X diesel
	(735 kW each)
Speed:	36 kn
Passengers:	114

8. The catamaran boom

The Polynesians were the first people to build multi-hulled watercraft. They were forced to do so because the islands they inhabited only yielded a limited amount of building material. But these palm logs were not comparable in shape to the oak logs of European commercial forests. The construction of large single hulls was therefore not possible, so the necessary buoyancy was generated by several hulls.

The Polynesian catamaran was also able to sail quite fast, so that this people could gradually cross the entire Pacific as far as New Zealand.

The motivation to build fast commercial ferries as catamarans also had to do with economy, but it was not about saving building material. The choice was between the monohull and the multihull in terms of how to build the simplest and most efficient fast ferry. Monohulls, in order to remain stable, have to be quite wide. Thus, they have to be developed as half or full gliders, which require a considerable amount of power to overcome their own bow wave. Only then do they glide along relatively economically.

If you cut the monohull in half, stretch each half lengthwise, seal it and mount the whole thing on a platform, you get a catamaran. This generates the same buoyancy as a monohull, but can use long and slender hulls that have less water resistance. It is essential that the length of the waterline is as long as possible. So it is not even absolutely necessary for the catamaran to start planing. It can often reach the desired speed in this way. The engine power does not have to be as great as on a monohull.

Catamarans are more stable because they float on two hulls. So you can build a ship that can be wider than a monohull. After all, it can't capsize! This makes it possible to carry more passengers and cars, which take up a lot of volume as payloads but do not weigh too much. So you can set up a comfortable and spacious passenger deck on a catamaran without space problems. Many a low-cost airline could take this as an example ...

In the 1970s, when the disadvantages of the hydrofoil and the hovercraft became apparent, experts in Norway and elsewhere looked for cost-saving alternatives to these complicated vehicles. In Norway, the aim was to improve the infrastructure of the country with its branching fjords by means of a water bus network. This should function like a land-based bus network. Above all, buses have to be economical and reliable in operation. They are also needed in large numbers to achieve adequate network coverage.

The experience gained with hydrofoils showed that they were by no means universally applicable and also had too many technical failures. Saving weight was too much of a priority with these vehicles, to the detriment of reliability. In contrast, fast monohull ferries were easy to build and operate, but required too much fuel.

In 1973, the Norwegian shipyard company Westamoen Hydrofoils, which had initially earned its merits in the fast ferry industry with the licensed construction of Supramar hydrofoils, produced the new type of vessel „Westamaran 85". As an asymmetrical catamaran, it was based on the hull design of the Norwegian „Storm-class" torpedo boats, which themselves were by no means catamarans.

The designers had cut the torpedo boat hulls in half lengthwise on the drawing board, pulled them apart a little and placed them under a platform that accommodated a passenger cabin and wheelhouse.

Stability and sea-keeping improved considerably, while still benefiting from the hydrodynamically advantageous hulls of the torpedo boats. Since the hull curvature of the catamaran hulls could only be found on the outside of the ships, while on the inside there were only vertical walls, this shape was called „asymmetrical".

„asymmetrical". While the torpedo boats could reach up to 50 knots, the „Westamaran 85" was content with a moderate 25 to 28 knots. The propulsion system was conventional with two 12-cylinder Mercedes-Benz diesel engines, reversing gears and two fixed propellers.

The design was immediately accepted by the market: by 1979, no less than 33 „Westamarans" of types 86, 88, 95 and 100 had been built (the number refers to the length in feet).

This economical and reliable fast ferry paved the way for the next generation of catamarans. Now, at last, a fast

seagoing vessel had been invented that did not have to be operated and looked after only by specialists. It did not require special berths and did not pollute the environment with excessive noise emissions.

To this day, hardly any „Westamarans" have been scrapped. They are now mostly operated by small companies in southern countries and carefully maintained. Their easy handling and economy suits the needs of these small shipping companies.

However, their days are gradually numbered, as they hardly comply with current safety regulations and can no longer be retrofitted to do so. Parallel to this development, there has also been a trend in Australia to use catamarans for various purposes. On the one hand, the waters in the area of the large coral reefs are not manageable for boats with a large draught, just like the small harbours with their shallow water depths. On the other hand, there are often high swells on the coasts, because the Pacific can roar undisturbed against the coasts of the fifth continent.

Catamarans were therefore recognised early on in Australia as seaworthy craft that have a shallow draft and great speed. In some industries, such as lobster fishing, having a fast boat can be very important. Because whoever can offer his haul of marine animals in a harbour first

often fetches better prices than the stragglers. Going fast is one of the catamaran's strengths.

In Australia, aluminium is a cheap material. Once you have mastered the problems of welding aluminium, you can build light and fast boats from it. It is no wonder that in the 1970s it became common to build light passenger ferries and water taxis as catamarans. There were even companies in the early days of this Australian industry that still riveted the aluminium as in aircraft construction. But the advantages of welding quickly prevailed.

A Westamaran 85 in a Norwegian fjord. The almost vertical inner sides of the asymmetrical catamaran are clearly visible.

Length:	26.67 m
Width	9.02 m
Draught	1.20 m
Propulsion:	2 x diesel MTU 12 V 396 TC 62
Power:	2 x 880 kW.
Passengers	166
Speed:	25 - 28 knots

(Above and below) Public transport in Australia can often look like this: smaller catamarans reliably ferrying passengers to work in Sydney Bay or taking tourists to the natural wonders in the Barrier Reef. Often these boats are built in simple shipyards on the beach and launched with the tide.

80

Catamaran success in Norway

The „Kommandeuren" was the first in the long series of the „40m Flying Cat" from Fjellstrand AS of Norway The shapely design was able to hold its own on the market right from the start. It is one of the most produced designs of its kind. About 60 were built, 35 of them under licence outside Norway.

A relatively small and compact shipyard company in Norway began to specialise in building catamarans in the 1980s. Then called Kvaerner-Fjellstrand, it was located in the town of Fjellstrand and had a long tradition of building fishing vessels and other small conventional ships.

At the same time, Norway began to build an alternative water transport network that would harness the speed of modern fast ferries for public passenger transport. These subsidised programmes led to a boom in Norwegian shipyards.

Kvaerner-Fjellstrand had already developed a 38-metre catamaran type and built it several times. This already successful series proved itself in service, because they had understood how to combine high shipbuilding quality with proven components such as MTU diesel engines. Incidentally, almost all the examples of this first catamaran series are still in service today in various parts of the world. Around 1990, Fjellstrand went back to the drawing board and improved the design. It was lengthened to 40

metres and the previously angular functional superstructure gave way to a streamlined but still spacious superstructure. The bridge was fitted into the new shape like an aircraft cockpit.

The first catamaran of this type built, with the designation „40m Flying Cat", was the „Kommandeuren". She was equipped with the new MDS system at the same time. The MDS is a horizontal movable fin pair mounted on short fin stubs under the two foreships of the catamaran.

The MDS has the task of significantly dampening the ship's tendency to pitch and roll. It worked excellently on the „Commanders".

The catamaran was able to blithely chase through waves on its test runs completed in quite drastic sea states. And it did so in a way that one would not have thought possible. Together with the streamlining of the superstructure, on which the overcoming water slid off, a fast ferry was now available that was completely up to the rough conditions of the north.

In Fjellstrand, there was hardly a shortage of orders. The „Flying Cat 40m" was built more than 125 times, including under licence in Singapore, the Netherlands and other countries. There were several major and minor modifications of this design, but the original sold best.

Why? At Kvaerner-Fjellstrand, not only was a high-quality product created, but care was taken to provide advice and assistance to customers in the factory. Therefore, even „beginners" in this technically difficult industry could quickly build up a reliably functioning ferry business.

The reliability of these catamarans quickly ensured that many hydrofoils and SES vehicles were replaced. With the catamarans, you might not have been able to travel quite as fast as with the other devices, but at least you got to your destination reliably.

Since 1990, „Flying Cats" have been built in almost uninterrupted succession. In 2004, the last of its kind was finally built. This does not mean, however, that Fjellstrand AS, as the company has been called since its separation from the Finnish Kvaerner Group in 2002, has given up building catamarans. They have only adapted the design to new regulations and needs and increased the range of equipment options.

Today's Fjellstrand designs come with more massive superstructures and a larger accommodation space for passengers and luggage. Nevertheless, the basic principles of the design have not changed.

(Below) The German „Halunder Jet" represents the later construction of the Fjellstrand catamarans. The superstructures, which are more massive than in the past, offer more space for passengers. According to today's regulations, the bridge crew must have a 360° panoramic view. This meant that the streamlined fit of the bridge into the upper passenger deck had to be dispensed with. The „Halunder Jet" operated between Hamburg, Cuxhaven and Helgoland for a few years, but was replaced by a new catamaran.

Optimising this design was more difficult than with ordinary catamarans. The „Halunder Jet" had to cover most of its distance on the Lower Elbe. The shallower water depth and the avoidance of waves on the river course was a point to be considered.

(Above) Even in rough seas, a „40m Flying Cat" can hold its own. It can still go fast in wave heights where its occupants might have lost all sense of fun by now. This photo shows the „commanders" during test runs for the MDS system, which consists of two steering fins mounted under the bow. It is supposed to significantly reduce the catamaran's tendency to pitch. However, with the waves pictured, any fin system would be overwhelmed.

(Below) a typical layout of the Flying Cat. Inside, the comfort is just as good as in a modern wide-bodied aircraft.

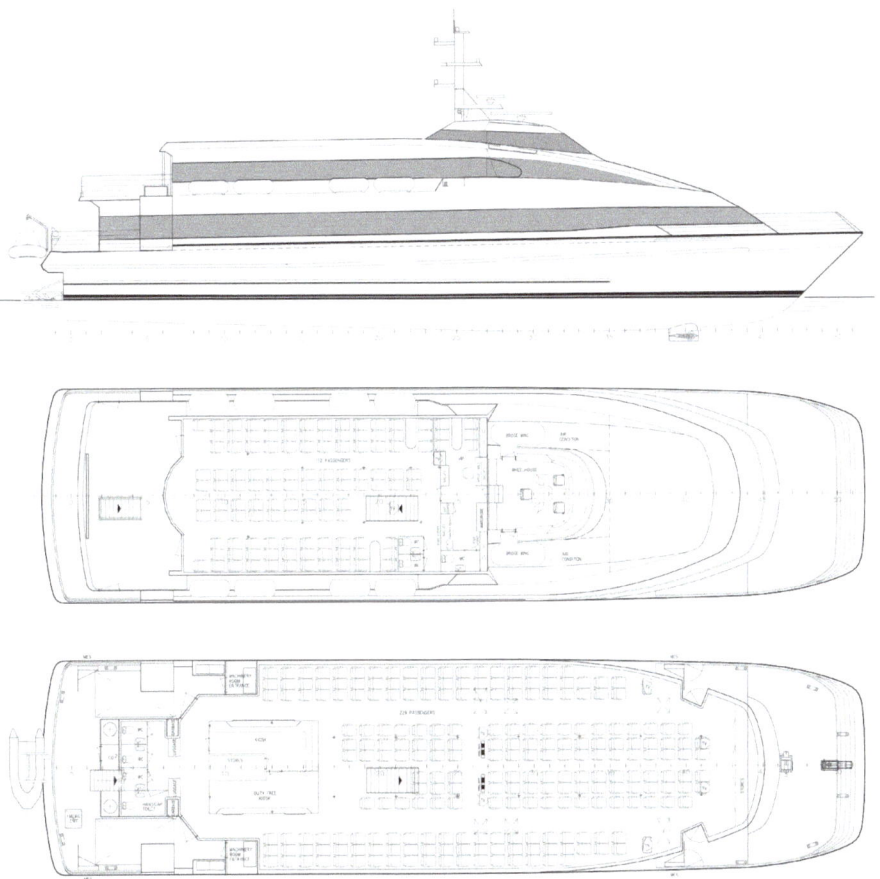

Crewboats - The bus to the drilling rig

In the early days of the offshore industry off the coast of Lousiana (USA), a new type of boat quickly became established, the crewboat. At that time, these were smaller boats, usually around 17 to 25 metres long. They were often mass-produced and were quite primitively equipped.

Powered by the noisy Detroit diesels often used at the time, they reached speeds of 17 to 25 knots. Their job was to cost-effectively ferry drilling crew replacements to offshore platforms, which were fairly close to shore at the time. The „bosses" of the oil companies had realised that you shouldn't put a helicopter in motion for every „roughneck", because they cost a lot of dollars even then.

It was a business for small shipping companies and nobody asked for permits, safety certificates or other „paperwork" for long in the 60s. This was also tried at the beginning of the oil boom in the North Sea, but the light boats proved to be totally overtaxed in this tough sea. Here the helicopter won.

Crewboats, like everything else, got bigger over time, sometimes reaching considerable sizes and becoming very powerful. They are used mainly in the more southerly waters and play an important part in supplying offshore installations. A typical Sikorsky S-92 offshore helicopter, which can carry up to 22 passengers, costs at least US$4,500 per flight hour. This is equivalent to the daily rate of a crew boat, which is slower but can also carry a lot of cargo in addition to passengers. Even in the oil industry, which is actually very strong in terms of turnover, calculations are hard today.

Crewboats handle the fast, light transports in many oil production areas.

(Below) The „Candy Counter", built in 2007, slowly manoeuvres towards an oil rig. The boat is 51 metres long and 9 metres wide. It has two diesel engines with 1,350 hp each and can reach 28 knots. It carries up to 60 passengers and 193 tonnes of deck cargo.

She is typical for her class. The passengers are stowed in the deck superstructure at the front on airline seats and are usually loaded onto the rig by crane via passenger baskets - so-called „widow makers". This is a rather dangerous manoeuvre, which is why more and more electronically stabilised gangways are used today. Such boats are often given less serious-sounding names - that's just the often sarcastic humour of the oil people.

The „Muslim Magoayev" and her sister ship „Rashid Behbudov" are currently the largest of all fast crew boats and were launched in 2015 and 2016 by the Australian shipyards Incat Tasmania in Hobart (Tasmania) and Austal Ships (Perth) respectively. Since then, they have been sailing in the Caspian Sea for an Azerbaijani shipping company. With a length of exactly 70 metres, these catamarans are longer than many a slow oil rig supply vessel.

Nothing is lacking on these ships: stabilising fins under the foreships ensure smooth sailing in heavy seas, there is a so-called DP system (dynamic positioning), which ensures reliable „hovering" over a predefined point by means of the jet drives and additional bow thrusters. This is especially important when approaching a drilling platform.

For as useful as the powerful drives may be when travelling fast at up to 38 knots, they make the ship difficult to handle at extremely low speeds. Theoretically, a small careless jerk on the thrust lever could already lead to a collision.

But the DP system, through a joystick and its programmability, offers the possibility to make the ship's reactions turn out exactly as needed. The two ships are certainly the biggest and best crewboats in the world at the moment.

Only a well-heeled oil industry can afford these investments. It is a fitting testament to performance for the Australian shipbuilding industry, which has already launched a number of world record-breaking vessels.

Technical data:

Length:	70.0 m
Width:	16.0 m
Draught:	2.0 m
Passengers:	150
Crew:	14
Deck area:	275 m²
Load on deck:	130 t
Cruising speed:	30 kn
Maximum speed:	38 kn

High-speed boats in use in the wind power industry

When the first offshore wind farms were built in Europe, intensive consideration was given to housing the maintenance technicians in large residential ships or on fixed platforms in the wind farm. Wind turbines that are defective do not produce electricity and must be repaired as quickly as possible.

On land, a van with several technicians drives to the shore and often only small, mostly electronic problems are fixed so that the turbine can turn again.

At sea, this is more difficult, but not unsolvable. A new type of vessel, the „Service Offshore Vessel" (SOV), quickly emerged in Britain, Denmark and other European countries. SOVs are fast small boats from about 15 metres in length that travel at quite high speed for several hours to a wind farm and drop off technicians on a wind turbine. Usually the repair only takes a few hours, so that one can go home again the same day.

A business model quickly developed for smaller shipping companies, which had often previously sailed tourists or earned a living from fishing.

At first, the vessels, mostly designed as catamarans, were built in small shipyards, but later shipyard groups such as Damen Shipyards from the Netherlands entered the business and sold off-the-shelf standard boat types together with cheap financing.

The boom in this business is now over and the remaining operators have become highly professional. Safety is the top priority. Only workers trained in special safety courses are allowed on board. The boats are fitted with special devices on the bow to allow them to dock without damage at one of the access ladders at the wind turbines. Large SOVs also have electronically stabilised gangways.

SOVs represent a new distinct genre of fast craft. They differ from the already known crewboats mainly in their smaller size. The reason for this is, of course, first of all that only small teams of up to four people are usually needed to repair a wind turbine. In most cases, 15 to thirty seats for passengers on board the boat are quite sufficient. Docking on wind turbines is a difficult matter. The boats are small and light and are effortlessly lifted by the swell or sink into a wave trough accordingly. In the North Sea, this can be a swell of at least one metre even in summer. So the technicians have to wait at the bow of the boat until the boat is at the highest point of the wave before climbing the ladder. Then they climb up as quickly as possible.

Some larger boats have electronically stabilised gangways, but it has been realised that these are too expensive and too complicated to operate. Such gangways are therefore now used more by larger offshore vessels.

The SOV's voyages are usually planned as day operations, because there are no sleeping berths for passengers. Also, service technicians are often not seafarers and are rather sensitive to swell and discomfort.

SOVs have only helicopters as competitors, but these are limited in payload and flight duration. So there is still room for both types of transport.

(Below) This smaller catamaran pins itself to the ladder using its powerful engines. Rubber buffers provide protection against damage.

SOVs exist in various forms. Above is a catamaran of about 20 metres built by AUSTAL, below a trimaran from the same manufacturer. There are SWATH, SES and also hybrid forms. The industry has been very creative here.

The introduction of the new „Westamaran" seemed like a salvation for the industry. Now the construction of catamarans, which were initially intended for passenger traffic, began worldwide. The use of aluminium became the decisive factor.

Aluminium is a very light metal that is extracted from bauxite ore, a type of clay. This requires a large amount of electrical energy. Some of the largest deposits are in Australia, which began to build up its own aluminium smelting industry in the period after World War II.

The metal obtained from the electrically heated smelting process is initially quite soft, brittle and very strongly oxidising. It would also be extremely unsuitable for use in salt water, as it would give off electric current like an alkaline battery when combined with other metals.

Aluminium is made into duraluminium by adding certain other metals such as manganese, which is mainly used in aircraft construction. The alloying process makes it a harder and more elastic material.

However, it still does not become significantly heavier than pure aluminium. A few more modifications to the alloy made it into so-called marine aluminium, which is very durable and yet light. One cubic centimetre of aluminium weighs about 2.3 grams, whereas one cubic centimetre of steel weighs 7.5 grams.

The advantage of the weight saving, however, is bought by the fact that the mechanical strength of the metal is significantly lower than that of shipbuilding steel. But first of all, the problem of processing the aluminium sheets had to be solved. All earlier large aluminium structures such as aircraft fuselages were largely held together by rivets or screws. Riveting had become easier with more modern tools such as the riveter and resulted in usable and durable structures.

For example, on the early large and successful passenger aircraft types like the Boeing 707 or the Douglas DC-8, not a single aluminium part is welded to another. In fact, only rivets, bolts and adhesives hold the whole „stuff" together.

The hydrofoils and hovercrafts of the 1950s and 60s were also riveted, but in the construction of the new catamarans, previously unknown joining methods were also taken up.

Aluminium can be welded with positive results if the weld is shielded from oxygen at the moment it is heated. This is done today with most welding equipment by blowing an inert gas such as argon out of a nozzle at the welding electrode. This allows the metal to join without a disturbing oxide film forming between the parts. With this, a method had been developed

of joining long pieces of sheet metal together. In fact, there were catamarans in the 1970s and 80s that were still held together by rivets, but these were exotic. The continuing construction boom in aluminium catamarans in Australia and Europe suddenly led to a shortage of trained aluminium welders in the mid-1990s. As a result, some

(Above) The success of a shipyard can lead to the ships to be built becoming larger and larger. In these cases, ingenuity is needed to still be able to use the shipyard halls. Apparently AUSTAL in Australia had this problem with the very large trimaran. But since the weather is often very good in Southwest Australia, building can also be done outdoors.

larger shipyards set up their own schools for this specialised metal profession to meet their own needs.

Aluminium is now the most widely used material in fast ferry construction. Designers have long since been able to compensate for its lower load-bearing capacity by the targeted development of special frame and stringer systems compared to steel. Problems from the early days, such as those caused by welding seams becoming brittle and material fatigue, have long been a thing of the past. Today, a modern large catamaran has at least the same service life as a modern steel freighter.

A major problem for designers and approval societies alike was the torsional stiffness of the new catamarans. Not only did they have to be built from a relatively soft and brittle material, but they also had to be able to withstand the different forces to which two hulls driven side by side by the waves are subjected.

The problem of torsional stiffness was solved by not simply attaching the two hulls to a thin platform, but by designing them statically as a solid, relatively torsionally stiff box. The box and the fuselages thus formed simple

The box and the hulls thus formed simple, independent static constructs that only had to be connected to each other. In addition, a very strong structure could be created by placing frames and cross struts („stringers") close together. After initial doubts, complex and large aluminium structures were increasingly recognised. Thus, different statements circulated at specialist conferences about how large the fast ferries made of aluminium may still become until the technical limits of the raw material are reached.

As always in such cases, this limit was constantly extended upwards, because the manufacturers seemed somehow unwilling to adhere to it. At present, it is believed that large aluminium fast ferries can only be produced up to a length of about 130 metres, but it is only a matter of time before even larger projects are tackled.

(Below) This trimaran is the shell of the USS „Independence", the „Littorial Combat Ship" (LCS). A grid of frames and longitudinal struts supports the hull. The building material is so-called marine aluminium, which is stronger than pure aluminium. It can only be welded under inert gas. In the past, aluminium ships were riveted together, but that is long gone. On the other hand, aeroplanes are still not welded.

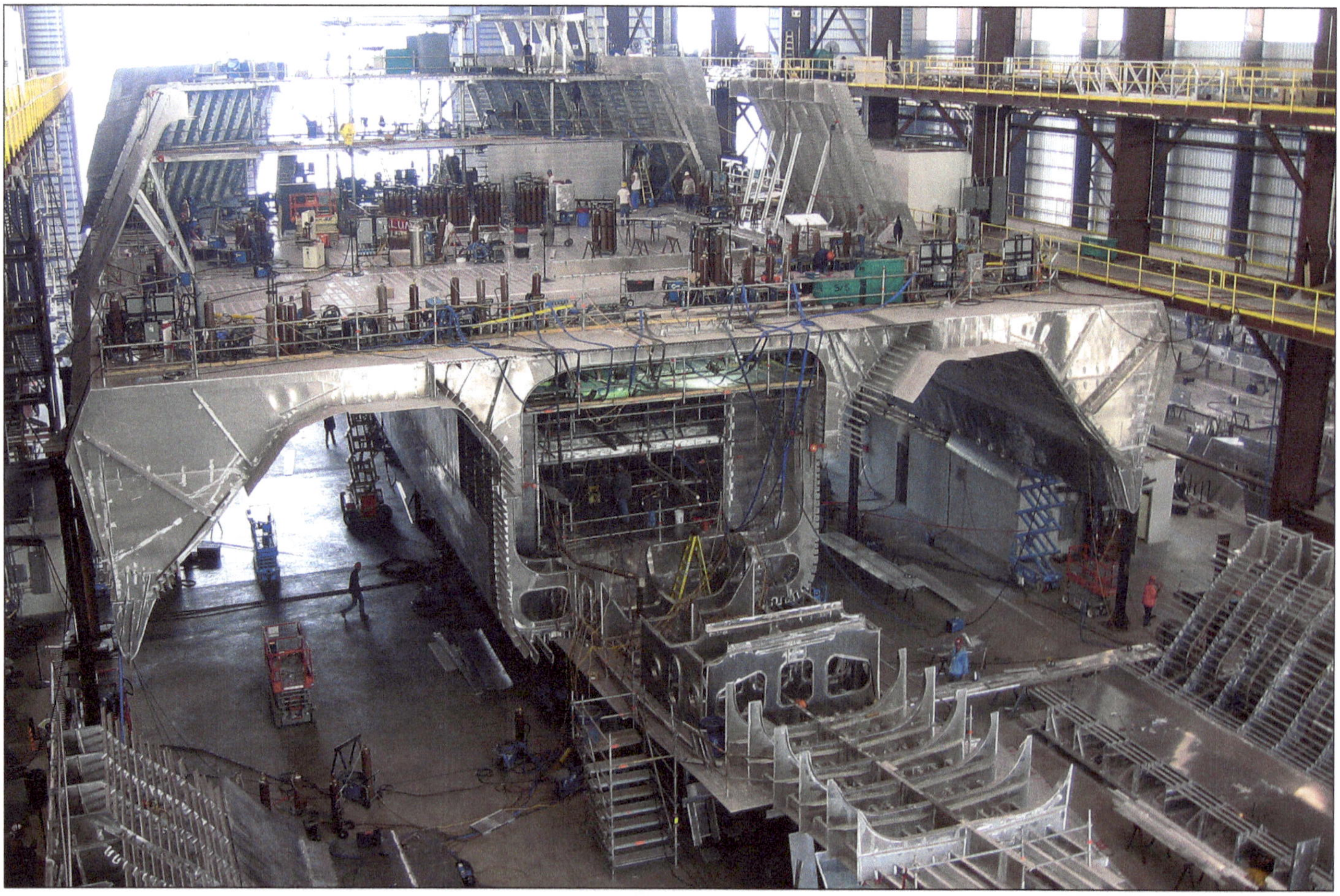

Like a rocket

With the catamaran building boom, there was at the same time a considerable upswing for manufacturers of urgently needed components, such as the propulsion systems and the engines. Today, fast ferries are mostly propelled by water jet drives or shaft propellers.

The discussion about which type of propulsion makes more sense is as old as fast ferry construction. The answer of the designers, however, is always: the drive that has been best matched to the hull and the operating environment is the best.

Decision-making on which system to use in a new-build project involves studies of the waters to be navigated, the water depths, the planned cruising speeds, the operator's expectations of the vessel's manoeuvring behaviour and many other details. However, there is no golden rule for the decision.

Wave propellers and jets have very different characteristics. The operation of the water jet becomes clear when you try to hold a fire brigade water hose on its own at full pressure. The recoil is so high that one person alone often cannot hold the hose. Newton's theorem that an action always causes a reaction, however, does not only apply to fire brigades.

In principle, every motorised vessel is propelled by the recoil of its propeller system or water jet propulsion. The old-fashioned idea that a ship's propeller screws through the water like a wood screw through material is simply wrong from a physical point of view.

The recoil of the almost 15-metre-diameter propeller of a large tanker is enormous, because with such a large circular area a huge amount of water is moved. However, the effect is based on the increase in flow velocity that the propeller imparts to the water. Admittedly, the thrust jet of the tanker does not look like that of a rocket, but the principle of action is the same.

More catchy is the way the water jet propulsion, known in the industry as „jet propulsion", works. With it, water is ejected through a nozzle located above the waterline with such force that it really does resemble the thrust jet of a rocket.

With the jet, water is sucked into a tube through a usually elongated oval opening in the bottom of the ship. In this tube, a so-called „impeller" is mounted at a widened point - actually nothing more than a propeller. The propulsion engine now turns this impeller powerfully so that the water sucked in is accelerated considerably and causes a strong recoil. Jets often have a system of deflection and control flaps at their discharge opening and the nozzle can also usually be swivelled from port to starboard. By purposefully deflecting the water jet, the recoil can also act to the sides or as a reverse thrust to the front. This means that the ship can now be steered without the need for a rudder blade. This capability is one of the reasons why jet propulsion is so successful. There is no need for a heavy reversing gear on the engine, and the construction of a vulnerable controllable pitch propeller system is also avoided.

(Right) A KaMeWa water jet for large fast ferries. KaMeWa, formerly a Finnish manufacturer, is now part of the Rolls & Royce Group.

The bracket on the thrust nozzle is intended as a swivel arm for the reverse thrust flap, which can be seen as a bulge under the nozzle's superstructure.

(Above) The jet drives develop a characteristic fountain when in use. It occurs because the jet has been mounted directly at the waterline or just above it.

Jet drives require much more power input at speeds up to 25 knots compared to a propeller drive. But beyond this limit, their efficiency becomes much better. The vast majority of high-speed ferries, which can reach at least 30 knots

are powered by jets. Jets vary greatly in size. They come with a nozzle diameter of only a few centimetres to a monster that could drive a truck through its nozzle opening.

The biggest enemy of jet propulsion is cavitation. This refers to a phenomenon in which small vapour bubbles form in the water at points of the jet propulsion where the flow velocity is particularly high.

There, the static pressure of the water becomes so low that water literally boils - as you can read in any physics book. Cavitation often occurs in periodic oscillations that can cause a ship to vibrate quite considerably. Another

(Above) The propeller of a waterjet is called an „impeller". Due to the flow conditions in a tube, the shape of the impeller blades differs considerably from those of a ship's propeller.

effect is the roughening of the walls of the water jet. This increases the water resistance inside the jet tube more than average and makes it unusable. Thus, in developing the very large jet drives, it was not simply a matter of scaling up a smaller design. Instead, the hydrostatic behaviour of these drives had to be tested at length.

For captains, jet drives are very convenient, especially on catamarans, because they make mooring a ferry much easier. Not only can a jet, whose water jet can be steered back and forth by the steering surfaces and reacts very quickly to commands, but it can even move the ship sideways. This is done, for example, by directing the left jet inwards and giving backward thrust, while its counterpart on the right side is also directed inwards but pushes forward. The sum of the force vectors results in a moment that moves the ship sideways.

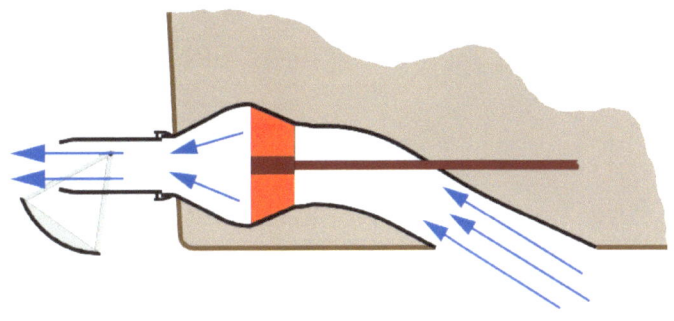

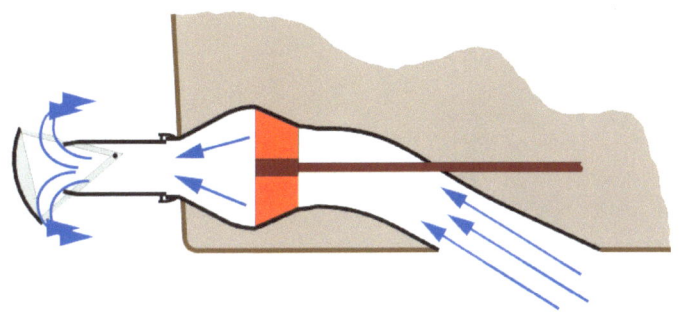

Above: The operating principle of all jet propulsion systems: Water is sucked in through an opening in the bottom of the ship and accelerated by the impeller (red). The jet exits through the nozzle, which can be swivelled to either side. The deflector flap (grey-black) is inactive during forward travel (but is directed in front of the jet stream to generate the reverse thrust. This allows a jet-powered vessel to reverse direction very quickly.

(Left) Jet units of type Rolls-Royce Kamewa 90SII at the stern of a speedboat. The output is 5,520 kW per jet and engine. They propel the Finnish Rauma-class boat to over 30 knots.

(Above) Four of the two „power packs" from MTU's 8000 engine series shown here fire this ship. Accommodating even such compact propulsion units is always a challenge for the designers of a fast ferry. Strong heat dissipation, vibrations, the mass of the engines and also the transmission of the generated thrust forces to the hull have to be taken into account. In a large fast ferry, the propulsion systems often account for up to one third of the total construction costs. Only 30 years ago, the construction of powerful yet compact diesel propulsion systems was not possible. Each of the engines shown here has an output of 8,200 kW or 10,996 hp. The engines have 20 cylinders and weigh about 48 tonnes including cooling water and engine oil (also known as „wet weight"). Consumption is about 1,885 litres per hour. The nominal speed is 1,150 revolutions per minute.

Whatever propulsion system a ship is equipped with, all systems require a strong power source. With a few exceptions, almost all fast ferries are powered by high-speed diesel engines. The „ancestors" of these diesels once stood in the engine rooms of diesel locomotives and speedboats. During their development into the standard engines used today, they have had to change very little. In the power range between 1,000 and 2,200 kW (1,341 to 2,950 hp), the southern German manufacturer MTU from Friedrichshafen has long since attained a certain leading position.

For a long time, the MTU 396 series in particular was a standard engine with 12 or optionally 16 cylinders that could be found in the engine room of a passenger ferry anywhere in the world. For the very large engines for the large catamarans, there was a tie between Ruston, Caterpillar and MTU. These bolides usually had between 16 and 20 cylinders and were in the power classes of

from 7,000 to 9,500 kW (9,387 to 12,739 hp). Their engine speeds were between 1,000 and 1,500 rpm, which is why they are also called „high-speed".

If you consider that a large catamaran usually carries four of these power plants in its belly, the unleashing of these forces can already lead to spectacular experiences. Above all, the acceleration capacity of such ships is always worth experiencing. The soles of the feet of spectators on the quay wall often tickle when such a large

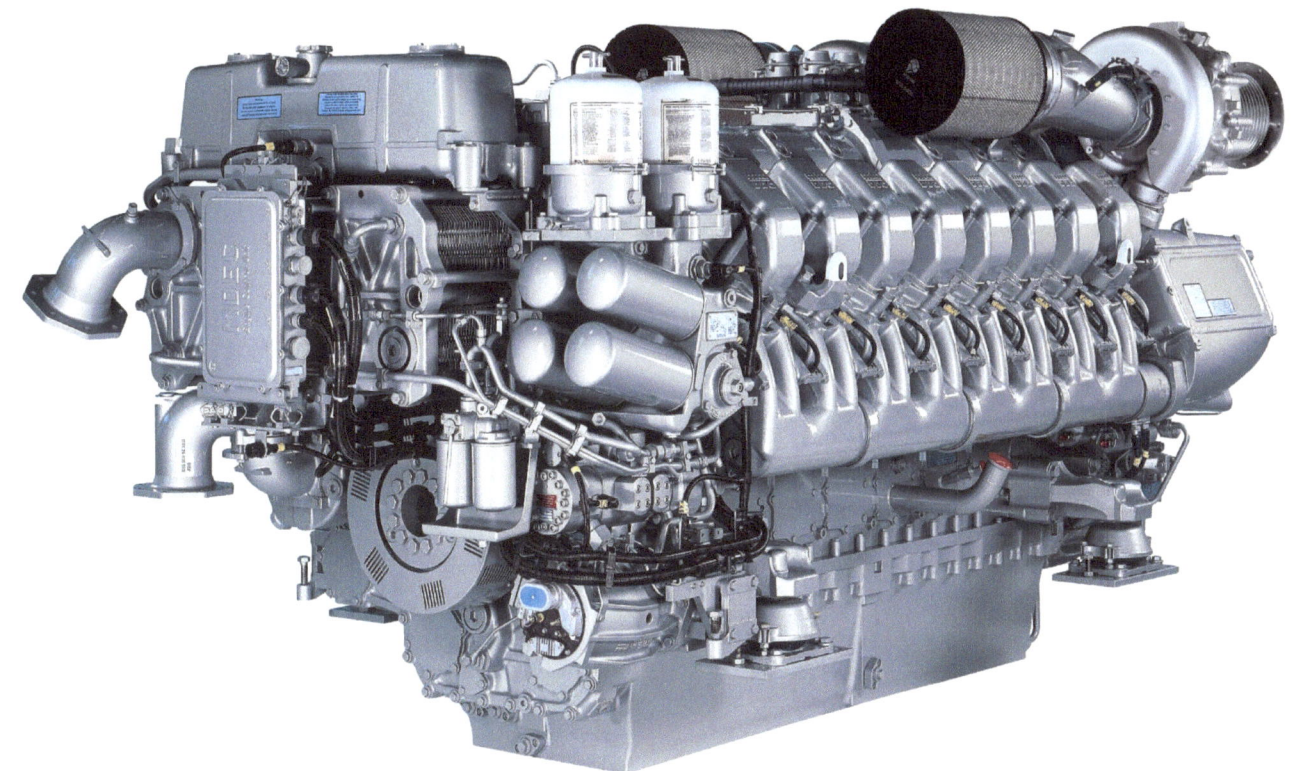

(Above) The 4000 series supplied by MTU has replaced the 396 series engines and comes with features such as common-rail injection and new engine electronics. Depending on the load factor and number of cylinders (8 - 16), power outputs of 1,000 to 3,800 kW are achieved. The load factor is determined for the planned mode of operation of the engine - depending on whether it is to be installed in a cargo ship, a motor yacht or a fast ferry. This makes the service life of the engine predictable until the general overhaul. A 16-cylinder 4000 series engine offered for installation in fast ferries weighs about 9.5 tonnes wet and delivers 2,465 kW or 3,306 hp. Engines of this type consume between 600 and 700 litres of diesel fuel per hour.

catamaran docks. The vibrations of the drives also propagate through the water.

There have also been some attempts to use gas turbines because of their very low weight combined with very high power output. Unfortunately, the specific fuel consumption of a gas turbine is much higher than that of a diesel engine. Also, many operators still shy away from the technology, which is little known to them. Only a few fast ferries have gas turbines. But they are often the fastest in their class, powered in this way. Gas turbines could well gain ground with the introduction of new fuels such as natural gas Characteristic of all diesel engines used in fast ferries is that they are usually loaded with almost 90 percent of their maximum power in continuous operation. This extreme continuous output causes them to wear out quickly and often leads to the engines having to be overhauled after only two or three years of ferry operation. This costs a lot of money.

Equally problematic is that these drives require a light and expensive diesel oil as fuel. This is two to three times more expensive than the heavy fuel oil so popular in „heavy" shipping. Only now are various manufacturers in the process of developing high-performance

for fast ferries that can also run on heavy fuel oil.

Heavy fuel oil is an almost tar-like black and often smelly fuel that actually comes off as a waste product of petroleum processing. But it was soon discovered that very slow-running and very large two-stroke diesel engines work excellently with this „dirt" and have a high degree of calorific value utilisation.

In fact, the heavy fuel oilers that are the standard type of propulsion on large cargo ships today are the most economical power engines available. They are huge machines, much larger than multi-storey houses, often weighing several thousand tonnes, with the crankshaft moving slowly at barely 100 revolutions per minute. The largest of them produce more than 80,000 hp. That is enough to propel a 300-metre-long container ship around the world at up to 25 knots. Only harnessing natural energy such as wind or solar power is more economical - not to mention protecting the climate.

Over time, manufacturers such as MAN, Wärtsila and

This engine room belongs to a modern In-cat catamaran ferry. It is equipped with two Ruston RK series diesel engines each. The robust RK engines have long been standard equipment at Incat. Today, Ruston belongs to the MAN/BW Group.

Ruston found out that smaller engines, which can only produce 2,000 to 8,000 hp, can also be powered by heavy fuel oil. These engines have crankshaft speeds of 400 to 800 rpm.

The shipowners thanked the manufacturers by placing many orders, because now a universal type of engine was available that could be installed in smaller freighters, ferries and on land even in combined heat and power plants. Nevertheless, the power-to-weight ratio of one of these so-called „medium-speed" engines is still several times that of a high-performance engine for light diesel oil, a „high-speed" engine.

Thus, only the largest fast ferries will certainly be able to benefit from the economic efficiency of heavy fuel oil. For only above a certain water displacement is the hull of a fast ferry capable of carrying several times the weight of a medium-speed ferry without losing too much in payload capacity. This size class is already ready for construction in the drawers of various design offices, but a market is also needed for „speeding giants" of this category. In the long term, oil and the products derived from it are substances that will no longer be available for use in engines in the foreseeable future.

What then should happen to shipping in general and fast ferries in particular?

Since a return to sailing on a broad scale is hardly to be expected and nuclear propulsion is also likely to be quite unpopular as an idea, forward-looking designers and economists are already looking into the question of alternative fuels.

Focusing on fast ferries, hydrogen and also natural gas are very interesting alternatives. Both fuels can be burned in diesel engines and in gas turbines. Cleaned up by catalytic converters, only water vapour and less CO_2 then flows out of the exhaust pipes. Already today, commercially used - although still slow - ships are operating in Norway that burn natural gas. The fuel cell, which in Germany is seen as the solution to many of the propulsion problems in land vehicles, currently has somewhat too little „power" for use in shipping. But development here will undoubtedly lead to systems that will deliver much more power than today's units. Compared to their heavy counterparts in conventional shipping, fast ferries have one major advantage.

advantage over their heavy counterparts in conventional shipping: they are

They are generally lighter because they are built of aluminium or plastic instead of steel.

Light ships therefore require less energy than heavy ones. Nevertheless, speed always exacts its price in the form of higher energy requirements. Therefore, fast ferries cannot be the solution to many maritime transport problems. Instead, there must be a transport demand on the routes travelled that justifies the energy input. The ideal environment for fast ferries are lines where they can significantly contribute to reducing a journey time compared to an alternative route on land.

(Above) The Rolls-Royce „Trent" is actually an aircraft engine that powers the Airbus A-380, for example. The core engine was removed for naval use. The „Trent" is intended for the almost 240-metre-long monohull fast freighter „FASTSHIP". Five of these engines are to accelerate the giant to more than 40 knots.

The „Trent" can deliver an output of up to 36,000 kW, making it the most powerful marine gas turbine currently on the market. The turbine alone weighs only 6.2, the complete installation container of the propulsion system about 22 tonnes. A comparably powerful diesel engine would weigh more than 1,000 tonnes. Its consumption is about 6,200 litres of diesel fuel per hour.

(Left) This Russian 42-cylinder diesel engine of radial design was an attempt to create a compact high-performance propulsion system for speedboats. The enormous number of cylinders required complex mechanics and made the engine a nightmare for mechanics. After all, it produced over 4,000 hp.

(Above) An MTU 8000 engine is lifted into an AUSTAL trimaran for assembly. The high-speed MTU marine diesels have a long history of development behind them. Originally, they powered large airships such as the „Graf Zeppelin" and the „Hindenburg". Light and powerful, they have been installed in speedboats since the Second World War. Today, MTU is a leader in the field of high-performance diesels for ships. Nevertheless, one can still recognise technical details that have not changed since the time of the airships.

10. The big fast ferries

![Fred. Olsen Express ferry]

(Above) AUSTAL's „Benchijiguar Express" was the largest trimaran built up to that time at 126 metres at the time of her delivery. But the following largest is only about three metres longer.

Size almost always impresses. If a large seagoing vessel can also travel particularly fast and looks futuristic, then it is sure to attract the public's attention. Huge high-speed catamarans, such as those in daily service today, were not even the subject of theoretical considerations only forty years ago.

No one would have believed that such a thing was even possible: ships over 100 metres long, made entirely of aluminium, powered only by diesel engines and racing through the water at 35 knots or more!

At that time, futurologists were still dreaming of nuclear-powered supertankers or passenger submarines. In shipping, steel ruled as a building material and the container was still in its infancy. Thus, the progress of the last 20 years can easily be seen by looking at prime examples. In the following chapters, several ships, ship series and product lines are presented that occupy a special place within the development history of fast ferry shipping.

They are representative of the overall progress that this industry has achieved.

(Below) The very large „Pacificat" was a flop as a project and caused a tangible political scandal in British Columbia (Canada) in 2000. The three 122 m long catamarans were supposed to power public ferry services in BC at 34 knots, but failed technically and commercially. They remain one of the few double-ender fast ferries to this day. They have never really been used, and are now laid up in Egypt. Eventually they will be scrapped.

The island of Tasmania in the south of Australia is actually known for its strange animal species like the Tasmanian devil and its particularly beautiful landscape. It is less expected that one of the world centres of the fast ferry industry is located there today. But the success story of INCAT Australia PTY Ltd. is associated with a high number of launches of large aluminium newbuildings and many record achievements.

INCAT's success story began with a disaster: in 1975, a large freighter rammed the Tasman Bridge, which spanned the long Hobart Bay near the city of the same name. The bridge collapsed in places.

The loss of the bridge made it impossible for many citizens to reach the other side of the bay as usual every day without major loss of time. So several local companies opened a ferry service with small passenger ferries to solve this traffic problem until the new Tasman Bridge was built.

One of these companies was the Sullivan Cove Ferry Company (SCFC), founded by Robert Clifford in 1972.

After about two years, the SCFC had carried more than nine million passengers across the bay, making a handsome fortune. The money was invested in building its own shipyard business, called International Catamarans PTY Ltd, which built a whole range of conventional small aluminium fast ferries.

But in the early 1980s, Phil Hercus, a ship designer from Sydney, had an idea of how to make the catamarans, which until then had been quite poor in heavy seas, more seaworthy.

He found an interested partner in Clifford, and both built the world's first wave piercer, the „Little Devil", which was only 8.7 metres long. The small boat proved itself, so they were soon able to tackle bigger things. With the „Spirit of Victoria", a 28-metre wave piercer was put into service in 1985.

The shape of this first wave piercer was strange: two flat catamaran hulls rising just above the water were connected to the platform by two thin streamline-shaped supports each.

This 96-metre Wave Piercer is called the „Evolution 10" class by Incat Australia PTY Ltd. As the first fast ferry in its size class, it was not built as a one-off, but is the result of a modular system of pre-planned equipment variants. Its individual components can be combined on the screen before construction until the customer is entirely satisfied.

(Above) This small ferry was the beginning. It made a lot of money during the reconstruction of the destroyed Tasman Bridge near Hobart. It was a slow catamaran, not a fast ferry.

so that the Anglo-American shipping group Sea Containers, which had also taken over the Hoverspeed canal ferries, among others, knocked on the door one day and was interested in the new design of a 74-metre-long passenger/car ferry.

Four were eventually ordered, to be used not only on the canal routes but also between Ireland and England. When the first „Hoverspeed Great Brittain" was launched, this catamaran represented the largest construction ever built entirely of aluminium up to that time. The new fast ferry could carry up to 850 passengers and 52 cars.

Passengers were accommodated in the fully glazed comfortable upper deck below the bridge. A complete novelty was that the passenger level was an independent module mounted on the hull with damping elements. This reduced the transmission of engine vibrations and noise to the passenger area. Car loading was either via the stern or via a bow hatch. Incat's new 74-metre class could cross the English Channel fully loaded at a cruising speed of about 34 knots, only 10 knots slower than the hovercraft.

The waves were now supposed to roar over the flat hulls without resistance and, if possible, not be impeded by the supports. The principle worked, but there was no damping against waves that came along at the same height as the catamaran platform. In addition, the supports were subjected to great loads and access to the engine rooms during the voyage was very difficult.

All in all, Hercus improved the concept considerably in numerous tank trials and finally arrived at the form that is still used today in INCAT newbuildings. The props merged into a slightly slanted sidewall and the platform got a new „false" bow. This normally did not dip into the water unless due to a wave that was too high. But instead of the wave crashing into the catamaran platform like a hammer, the false bow splits the wave and dampens its impact considerably. It also creates a further momentary lift on the forecastle to keep it out of further waves.

The passenger catamarans „Tassie Devil 2001" and „Quicksilver", built according to this shape, were able to cope with considerably worse weather than the previous catamarans with this aggressive-looking shape of the forecastle. Word of their success got around,

than the hovercrafts. The launch of the new catamarans was a great success. „Hoverspeed Great Brittain" won the „Blue Riband" on the 1989 crossing from New York to Southhampton.

After this success, a whole series of 74-metre catamarans were ordered from Incat Tasmania, which were used in the Baltic Sea, the Irish Sea and also in the Mediterranean. The large shipyard in Hobart developed into the clock line for fast ferries of the same type.

When a construction stage was finished, the hull was

(Right) „Little Devil" was the first wave piercer ever. It still stands today as a floor model in front of the Incat Designs Sydney headquarters in Australia.

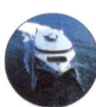

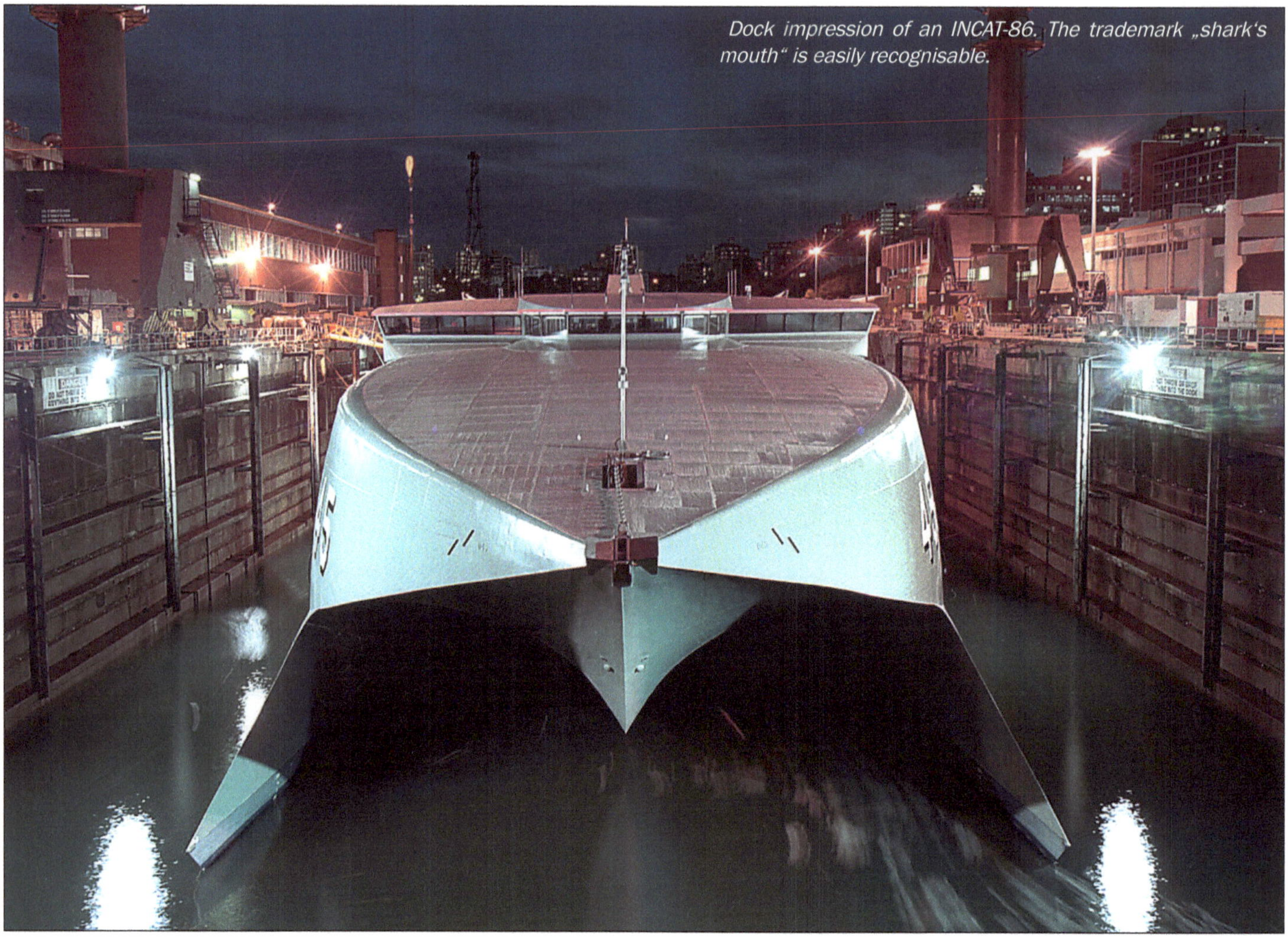

Dock impression of an INCAT-86. The trademark „shark's mouth" is easily recognisable.

pushed a little further towards the water, and the hull following behind it was further assembled to the same level. The construction department also had plenty to do. The 74-metre class was followed by the 76-, the 81- and the 86-metre class in fairly short succession. The bridge, which was initially located in front, slid midships, giving the ships the impression of having a kind of nose with wide-open jaws underneath. For sailors and surfers, the sight of an incat wave piercer often gives the impression of a gigantic predator hunting them down. This inspires quick evasion.

Incat began to build hulls in stock during the boom. These hulls were built up to a production level of 80 % without concrete orders. The now buying customer could therefore take delivery of a fast ferry within three months and start his ferry service, as the final equipment mostly consisted only of the installation of the interior fittings. This risky sales method led to Incat Tasmania being on the verge of insolvency in 2000 and only being able to maintain its solvency through a government loan.

Incat survived the crisis and introduced a completely new design with the 96/98-metre class, which outwardly appeared to be a continuation of the previous model series, but featured considerable innovations in many details. The vehicle deck of the last conventional classes, the 86 and the 91-metre generation, had a lane layout due to the constant enlargement of the ships. This allowed the simultaneous loading and unloading of the vessels, but even experienced marshals could not be confused.

The new class, named „Evolution 10", again had a simple superstructure and only one turning circle at the forecastle. Instead, several lanes could be raised by new lightweight lifting platforms so that two layers of cars could be accommodated on one lane, but the ship could still carry high trucks or buses. This made the new generation of ferries the most flexible of their kind, as competi-

tors' catamarans and monohulls had essentially rigid and unchangeable vehicle deck configurations.

The new 96-metre-long „Evolution-10" class and the „Evolution-10b" class, which was enlarged to 98 metres, were very successful. Two each went to Trasmediteria and Fred Olsen Lineas of Spain. The US Navy bought a ship of this class, the „USS Spearhead", as a special logistical support ship.

But then the next Incat generation was completed, the 112-metre class. This class, known as the „Evolution-One12", was built for a Japanese ferry line and planned for the US Navy to tender for. In the process, the hulls resembled each other up to a certain degree of completion.

This modular concept, called „Seaframe", makes it possible to offer customers the required components such as the

passenger area, the crew cabins, certain types of ramps or even a helicopter hangar. Each of the modules has already been completely designed and calculated in terms of safety and statics, so that considerable time can be saved because the ship does not have to be recalculated in the combination of modules chosen by the customer. As far as the propulsion system is concerned, Incat has so far stuck with its in-house supplier Ruston from England, which was taken over by MAN several years ago. Caterpillar and MTU have not had a chance so far.

Only the planned „Revolution 120" class was to introduce heavy oil propulsion as an absolute innovation. The use of heavy fuel oil will lead to considerable financial savings in the operation of such large fast ferries of almost 120 metres in length and 30 metres in width. However, heavy fuel oil is currently under fire because of its toxic exhaust gases (sulphur dioxide and NOX). Instead, LNG (natural gas liquefied by cold) is now being offered as a fuel. This would be burnt in gas turbines.

The latest concepts are now switching to batteries and liquid hydrogen. The lightweight construction of aluminium has great advantages, because it allows the use of less energy than a steel ship with the same payload and speed.

(Right) No generational problems: each new series built on the experience of the previous generation. From top to bottom:

- *74-metre class*
- *76-metre class*
- *86-metre class*
- *96-metre class (cargo version)*

103

(Above) This 96-metre wave piercer built for Fred Olsen Lineas in the Canary Islands initiated the leap forward in the development of Incat products. The internal space layout was made more flexible and easily adaptable to the customer's needs, especially on the vehicle deck. Also, for the first time, exhaust posts were introduced on the aft deck.

(Right) Approximately Incat Australia's current product line for this century:

Besides the new small 17-metre series, only the largest newbuildings are offered. The already impressively large „Evolution 10b" with a length of 98 metres (second from bottom) represents the lower end of the scale. Meanwhile, a catamaran of around 130 metres is offered.

(Second from top) The so-called „Fuel Miser" was designed as an economical freighter to operate as a truck ferry at only 18 to 24 knots slower than the fast ferries and to be much more economical to operate than conventional steel „flat irons" due to the Incat's usual design features.

The new Incats will also be offered to military users in a number of variants as transporters, mine hunters or fast logistics vessels.

However, Incat has lost a very important tender from the US Navy for the delivery of 12 so-called „Expeditionary Fast Transport (EPF)" ships. In future, these will transport vehicles, soldiers and goods at more than 30 knots across the world's oceans to increase the response capability of the armed forces.

A catamaran, for example, can carry hundreds of tonnes of cargo and cross the Atlantic in just three days. At the same time, it is much more economical than a comparable fleet of transport aircraft.

Incat now also offers catamarans for offshore use. The first of this kind is described in a later chapter.

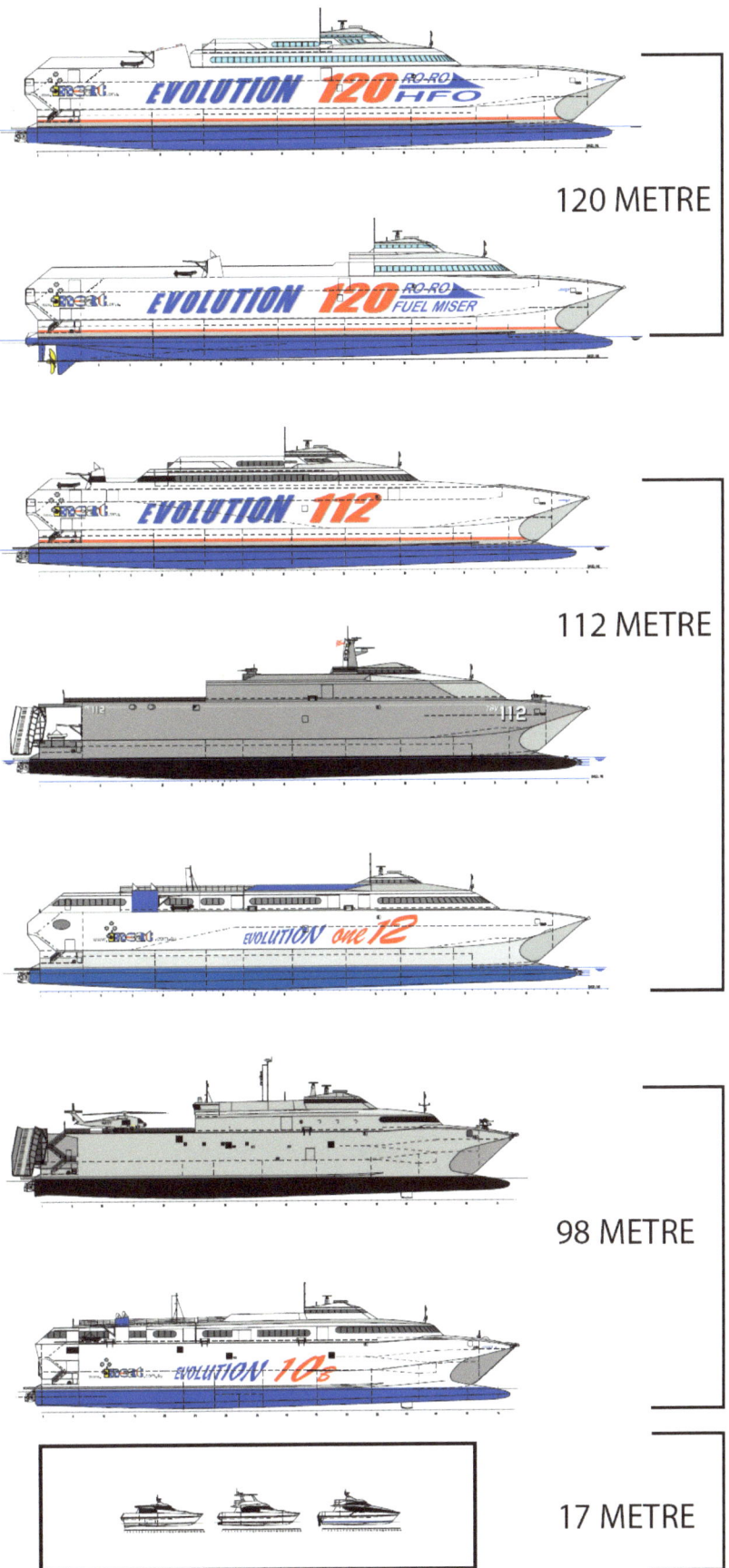

120 METRE

112 METRE

98 METRE

17 METRE

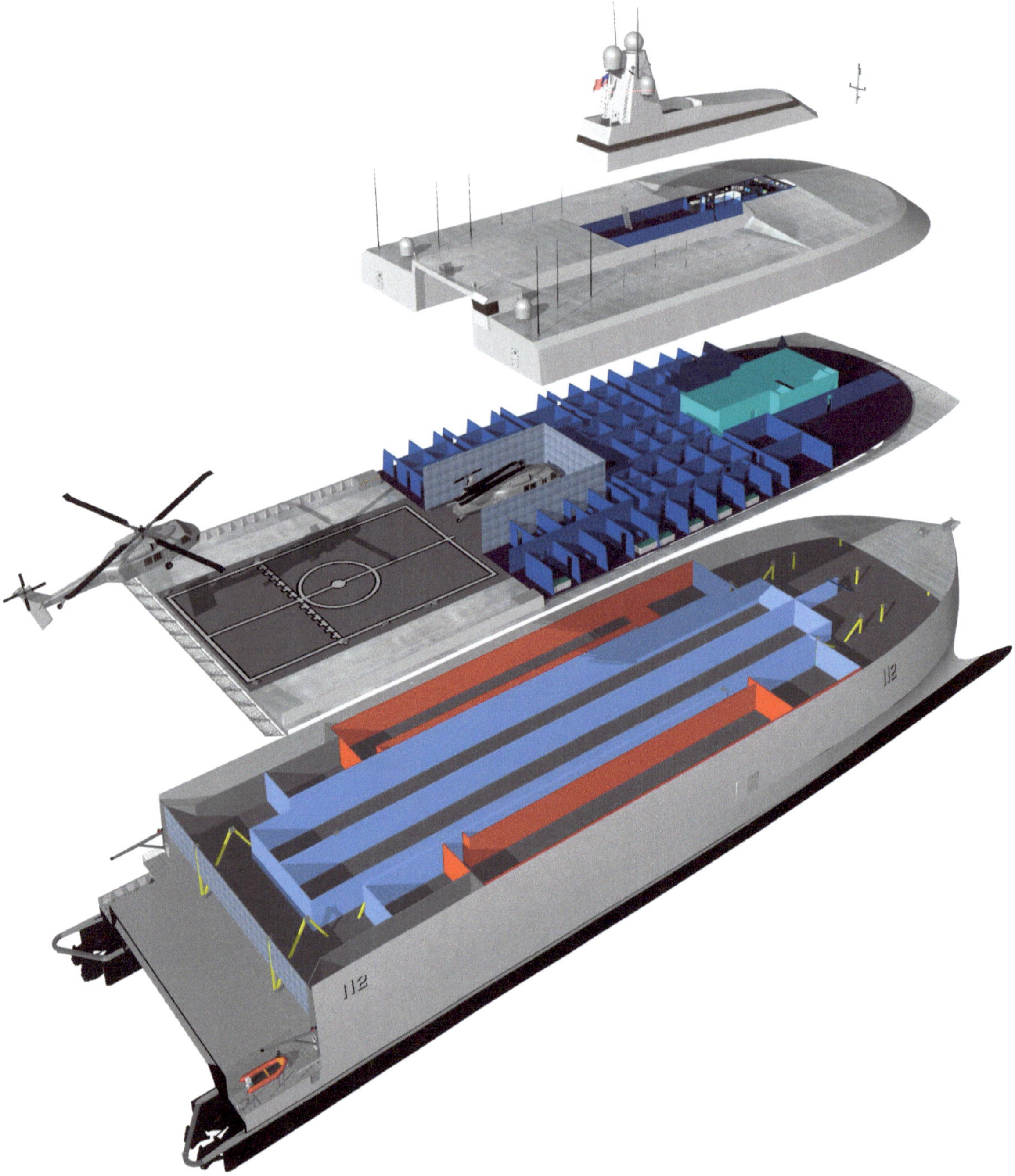

Above) The „Seaframe" concept developed by Incat allows the construction of a large number of variants on one and the same hull. The 3D rendering shows a cut-open EV-120 virtually outfitted as a military supply and transport vessel. This and similar variants were offered to the US military authorities.

During the tsunami disaster in the winter of 2005, a similar vessel, the „Spearhead", was dispatched across the Pacific as a fast emergency transporter with relief supplies. Such a large catamaran can also cross global distances at speeds of more than 30 knots. This represents a new dimension in military logistics.

Progress from Western Australia

AUSTAL Ships PTY Ltd. is a 25 year old shipyard company based in Henderson on the Western Australian coast. AUSTAL also started small with fishing vessels, government boats and small tourist catamarans. In contrast to INCAT, AUSTAL has maintained a wide range of products to this day, which has contributed to the great economic stability of the company. Today, the product range includes small and large fast ferries, patrol boats for the navy, small cruise ships and other vehicles.

Austal built relatively early on the concept of the semi-SWATH already described. A particular success was the construction of the new catamaran „Speeder" in 1991. The shapely ship with a length of 43 metres was capable of almost 44 knots.

The construction of the 82-metre ferry catamaran „Delphin" was celebrated as another great success. The „Delphin" could take 175 cars, 4 buses and 600 passengers on board. Powered by four MTU diesels with 20 cylinders each, she crossed the Baltic Sea between Strahlsund and Trelleborg for about 10 years. The development of oil prices and a change in the business policy of her operators prompted them to sell the „Delphin" in 2005.

The compact catamarans of the 82-metre class and their other counterparts, which differ only in dimensions, turned out to be „big sellers" on the market.

Operators from the Mediterranean, the Orient and many other parts of the world ordered this design by the dozen. One of the reasons was the cleverly planned interior design with a vehicle deck that had two raised decks on the sides for cars with ramps that could be raised. This allowed vehicles to be stowed on two lanes amidships with a large clearance height.

Above this was the passenger deck. The hull shape was suitable for sea areas with not too bad weather, but ferries of the 82-metre class usually had to cease operation from an average wave height of 2.5 to 3 metres. However, since the AUSTAL catamarans were somewhat lighter than incat ferries of the same size, they were able to operate in the same conditions.

Incat ferries, they could be operated more economically. It is noticeable that at the same time as AUSTAL

The busy scene from AUSTAL Ships' yard in Henderson shows a stable order situation. These are only the smaller ships that AUSTAL usually builds. The „big lumps" are manufactured next door. Over the years, AUSTAL has taken over several other catamaran builders such as International Catamarans and Wave Master at the Henderson site. The result is a versatile shipyard group focused on aluminium shipbuilding in various forms.

was successful with this class, Incat began to concentrate on building much larger ships.

So it can be assumed that these two leading manufacturers had divided the world market between them. As already described, AUSTAL came onto the market in 2004 with the new 127-metre trimaran. The very impressive ship excited the journalists and the public. In the meantime, there are three large civilian AUSTAL trimarans with lengths between 102 and 126 metres. AUSTAL succeeded in winning the competition for the construction of the new „Littorial Combat Ship" (LCS), a new type of fast combat ship with anti-radar capabilities for coastal operations, in the USA in cooperation with General Dynamics. The design submitted is a trimaran of the same size as the Fred Olsen Ferry.

The LCS occupies a key position in the US Navy's anti-terrorism strategy.

The success of this trimaran will undoubtedly in turn have some influence on the commercial shipping industry, which will then be more likely to accept this new hull form. AUSTAL, as a supplier of large and small fast ferries as well as a whole range of other ship types, is broadly enough positioned to be able to survive bad years. In this industry, bad years mainly occur when the world market price for oil rises to unusual heights.

But the real breakthrough for AUSTAL was the LCS for the US Navy and the logistics catamarans already mentioned. Today, AUSTAL is the largest aluminium shipbuilder in the world. From small craft only about 19 metres long to oversized trimarans and catamarans, AUSTAL has an enormous range of products to offer.

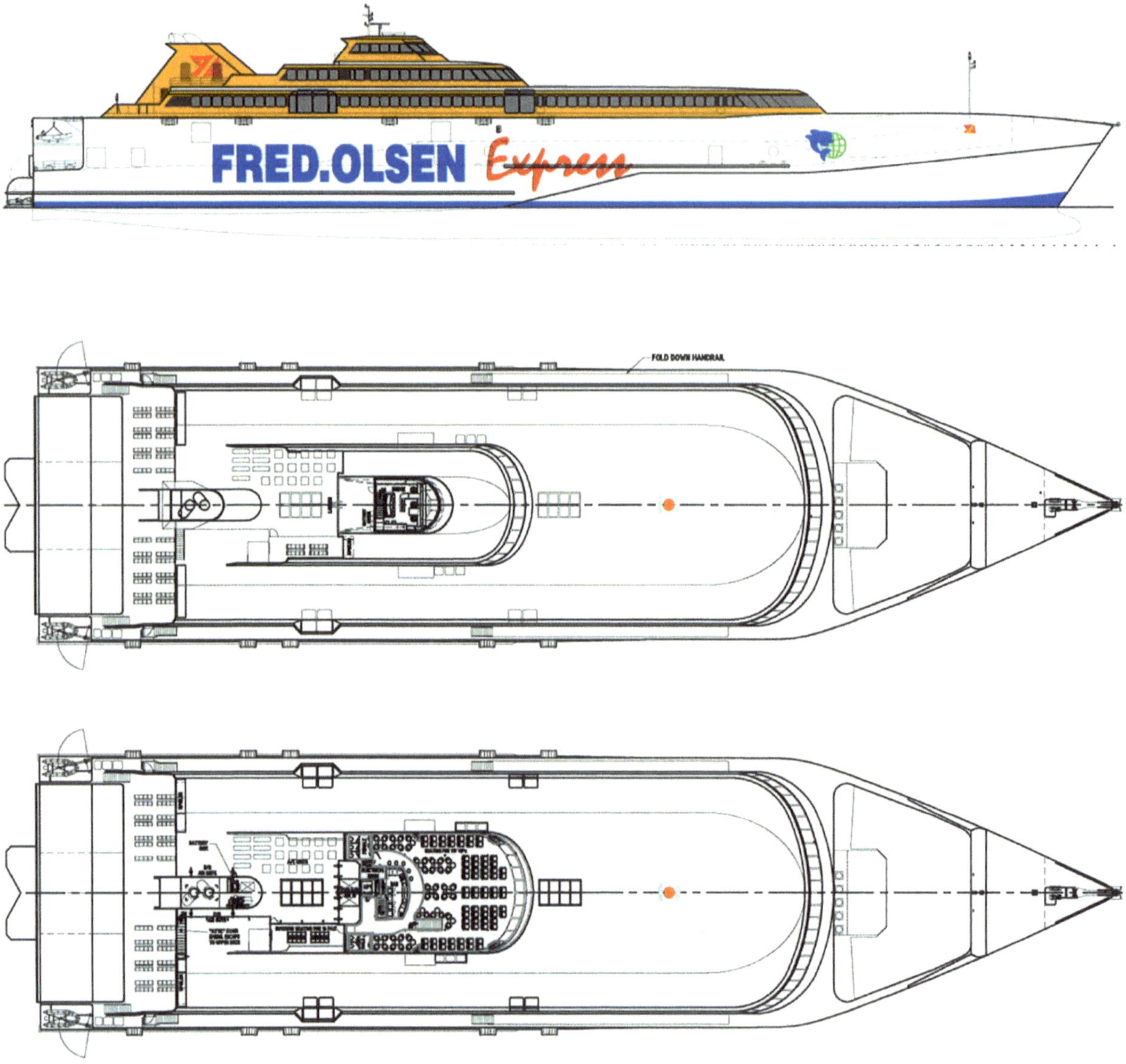

(Above/following page) The interior of the AUSTAL trimaran: „Benchijiguar Express" (126.70 m length, 40 knots, 1241 passengers and 341 cars).

108

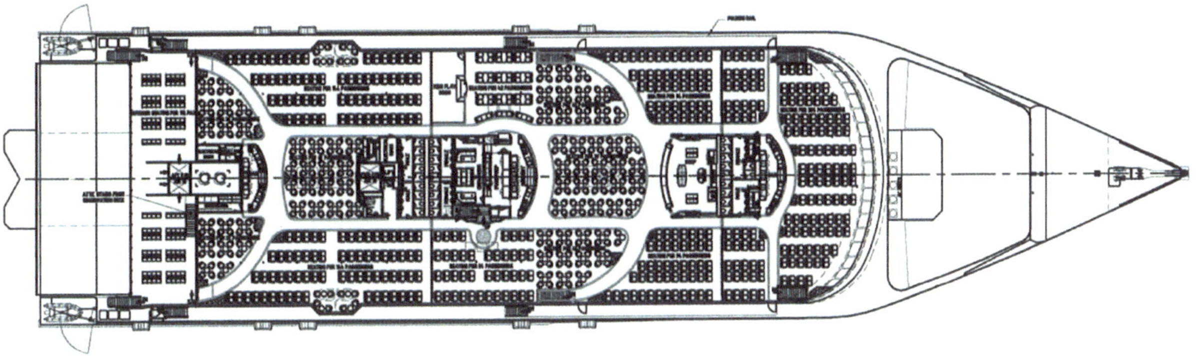

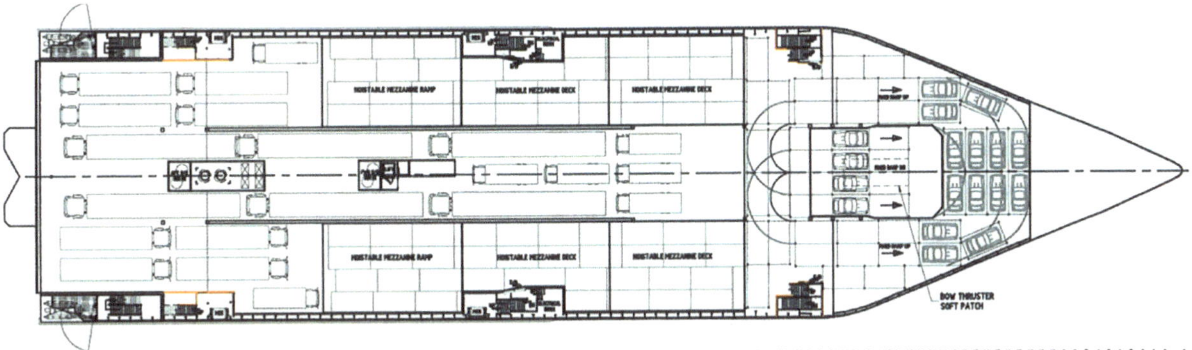

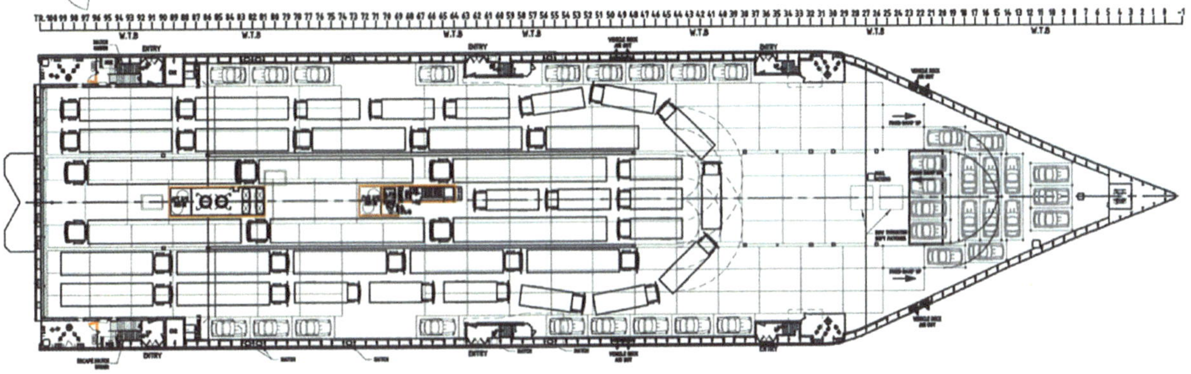

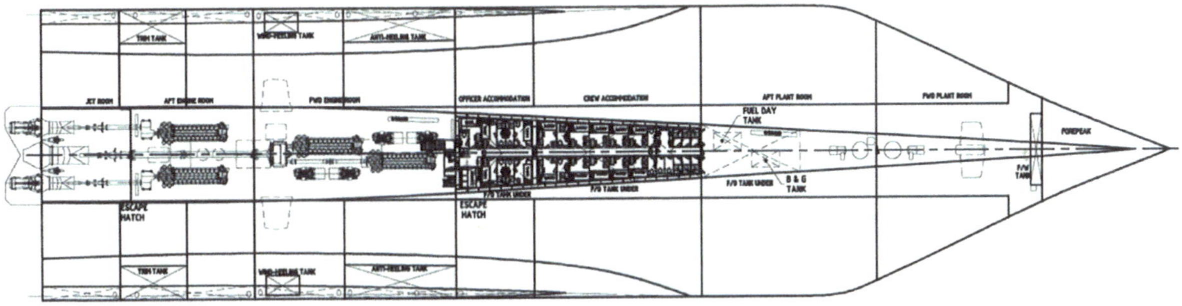

(Above) The US Navy's Littorial Combat Ship (LCS) has been specially designed for coastal operations. STEALTH capabilities, a new type of light armour, new surface-to-surface missiles, as well as special remote-controlled helicopter drones give it great assertiveness. In order for the LCS to achieve a speed of more than 40 knots, a hull form new to warships - the trimaran - had to be chosen. There is currently nothing comparable outside the USA.

(Below) Built in 2009, the „Superferry Hawai" is the largest catamaran AUSTAL has built to date, measuring just under 107 metres in length. It and its sister ship went to the US Navy under the names USNS „Guam" and USNS „Puerto Rico" after the operator went bankrupt in 2012. Today, they transport material and vehicles.

Japan's dream - The Super Technoliner

The largest sea-going vessel supported by air cushions - we remember the term „SES" - comes from Japan. The Far Eastern country is located in a sea area that offers very unfavourable initial conditions for the use of fast sea vehicles. At the same time, the country is divided into many islands that need to be connected. But not everywhere can bridges and tunnels be installed on the opulent scale usual in Japan.

In a country as technologically advanced as Japan, the development of new high-speed vessels such as hydrofoils was closely observed shortly after the beginning of the 1950s.

When the PT-20 and PT-50 boats, built under the direction of the Swiss Supramar AG, gained a foothold worldwide, the Hitachi Group acquired a licence to build them for use in Japan. Since Japan has its own strict licensing regulations, they could not simply import craft from abroad.

Later, Mitsui Heavy Industries developed its own passenger hovercraft. Other industrial groups were involved with monohulls and also with SWATH fast ferries.

In Japan, technical universities and institutes began research early on to create large, fast seagoing vessels. It was already clear that the smaller ferries would not offer sufficient capacity in the long term to satisfy the need of the inhabitants of distant islands for connections to the economic centres of the country. One of the main directions of research was the development of oversized SES vehicles. The reason for the interest in the SES concept was the high speed potential of these vehicles.

For although smaller-sized SES are subject to the „cobblestone effect", this no longer has such a serious impact on vehicles with lengths of 90 metres or more. On the other hand, the frictional resistance of large SES is lower than that of conventional fast ferries of the same size. However, the calculation seemed to work out, because the state subsidised the development of a large SES, which was to serve as a test object and at the same time be used commercially.

The ship was not planned as a conventional ferry for short distances, but as an island supplier for the Ogasa-

The TSL-A, christened „Osagawara", has an uncertain future. Its design as a cargo passenger liner with cargo holds instead of a vehicle deck makes it difficult to sell internationally. Only the military could use the expensive super-fast vehicle to boost their logistics. It is currently on.

wara Islands, which lie about 530 nautical miles (981 km) south of Tokyo. The journey at a service speed of 39 knots would take about 17 hours. Therefore, the TSL-A, which means.

„Techno-Superliner A" means, was equipped with passenger cabins and cargo holds. The TSL-A is currently the largest fast ferry - and also one of the fastest. It can reach about 44 knots at full speed.

In earthquake- and tsunami-prone Japan, the TSL-A was also intended for the important secondary function of a disaster relief vessel. Loaded with relief supplies, the „speeding giant" can reach any point on the Japanese coast in just a few hours, unlike land-based relief workers who may have to fight their way across destroyed roads and bridges.

To build the TSL-A, numerous new developments had to take place in the propulsion systems, the apron systems and also in the hull area. For example, the TSL-A has the two largest KaMeWa VLWJ 235 water jets ever built, which are also urgently needed to convert the power of the two General Electric LM 2500+ gas turbines, each with 25,180 kW or 33,766 hp. A jet of water ejected by such a jet is powerful enough to fill a very large swimming pool within fractions of a second. But no swimming pool could withstand the force of this water jet.

The TSL-A is powered by a total of 4 diesel engines, each producing 4,000 kW (5,364 hp). That is more power than most fast ferries use for propulsion. The 140-metre-long TSL-A carries 740 passengers and up to 210 tonnes of cargo in containers. The width is 29.8 metres. The dimensions and performance of this ship are certainly impressive,

but they also explain why even before the launch, around mid-2005, the intended charterer, the Ogasawara Kaiun, pulled out of the charter.

They were unable to operate the ship without incurring ruinous losses, because oil prices had risen enormously since development began in 1989. Thus, the dream of Japanese naval technologists was suddenly shattered and the Ministry of Transport was forced to develop a rescue plan. However, the enormous operating costs of the TSL-A are a major obstacle to this.

Today, hardly any material about the TSL-A can be found on the web. It almost seems as if people deliberately want to forget about it. This often happens in countries where people are not used to failures with such large projects.

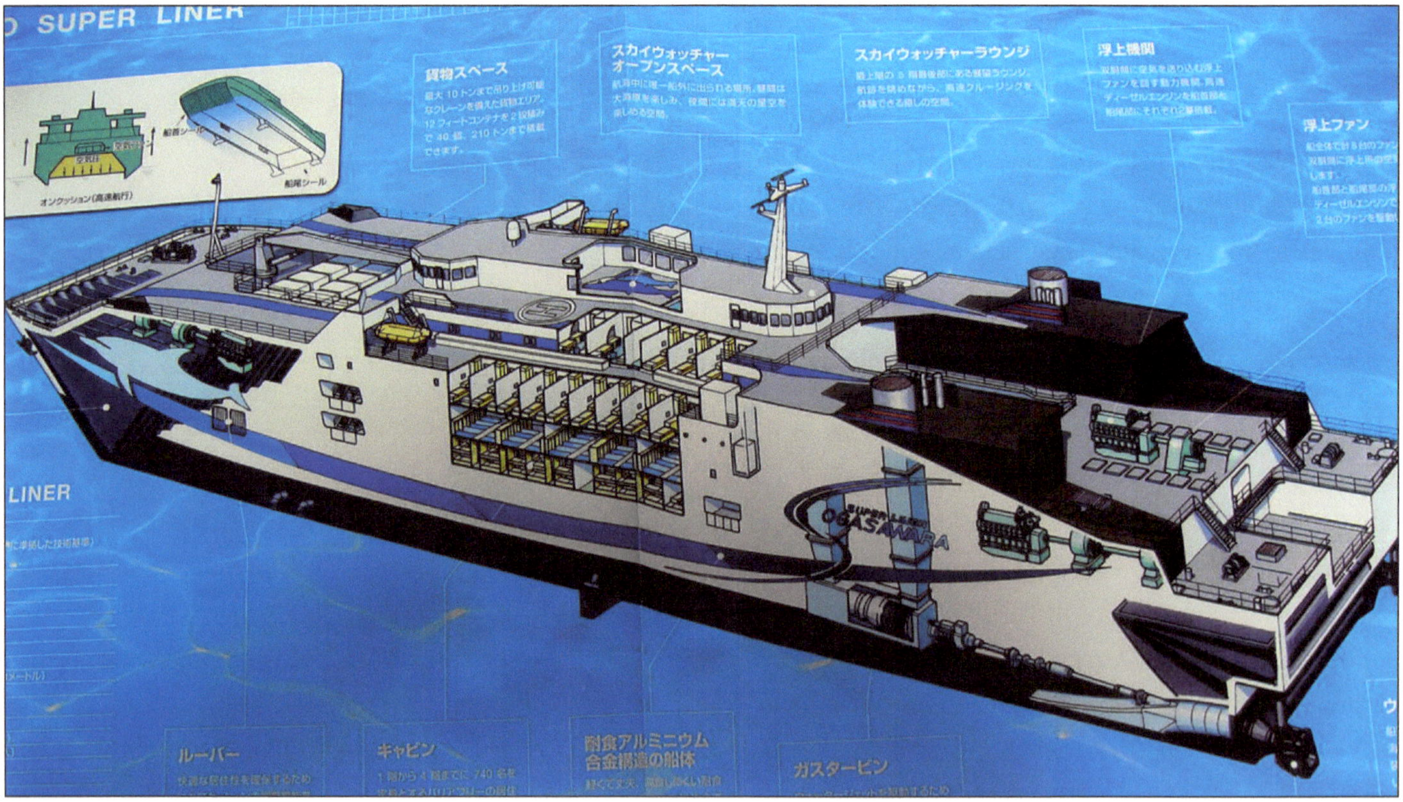

Pictures of the TSL-A layout are hard to find. This brochure was „captured" in Japan. It schematically depicts the basic structure of the ship. It is the largest hovercraft ever built.

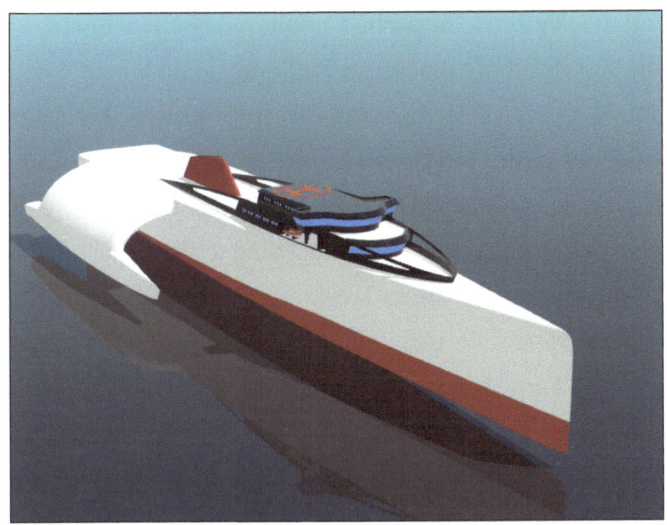

Length:	160 m
Width:	51 m
Draught:	5.15 m
Speed:	32 kn
Payload:	2,150 to
Truck trailer:	up to 94

France is a country where unusual technical innovations are welcomed and encouraged. People are proud of technical successes such as the Concorde, the world record-breaking TGV train and other modern achievements and do not demonise them, as is quite common in other parts of Europe.

Within the traditional French marine industry, the still young company BGV International from the south of France had developed very ambitious plans to speed up sea transport. The company's designers had been working for several years on a trimaran concept that would lead to a whole product line of innovative high-speed ferries.

The vessels consisted of a completely aero- and hydrodynamically shaped, long and slender hull that was more like that of an aircraft than a ship. On the sides were two wing-like supports on which small auxiliary hulls extended into the water.

The way in which the ship was to gain additional stability was also innovative. The connections between the auxiliary hulls and the main hull were actually intended to create aerodynamic lift to reduce the tendency to roll.

However, the new design was not a passenger ferry, but a RoRo freighter that would carry only truck trailers. BGV France, in cooperation with the port of Boulogne-sur-Mer, had been trying for several years to establish several ferry lines linking Oslo, Boulogne-sur-Mer and Vigo.

Cooperation had been agreed with a major maritime transport company. The BGV ferry, just under 160 metres long, was to travel these routes at around 30 to 35 knots.

An economical heavy oil drive would have saved considerably on transport costs. It is important to know that Boulogne-sur-Mer is one of the most important centres of the French fish processing industry. But in the meantime, the raw materials of this industry are no longer landed directly there, but brought in from all over Europe by truck.

The increased truck costs due to tolls, labour costs and fuel prices as well as taxes should be reduced again by using the BGV type of vessel.

However, the first customer, a company called „Chikara Shipping Ltd" turned out to be a scam. The investors lost their money in 2008 and the British initiator of Chikara Shipping fled abroad. Since then, the project is basically dead and will certainly not start again. That is a pity, because BGV would certainly have set a great example.

11. Flying ships - the airborne future?

Probably the most exotic concept for escaping the resistance of water is the so-called ground effect aircraft, also known as „WIG - Wings in Groundeffect". They are a hybrid of aircraft, flying boat and ship. After almost 50 years of research, this technology has reached a stage where it may soon be possible to realise this concept.

The ground effect has been known to large seabirds for a very long time and to humans since the time of the Second World War. When a flying object flies with its wings at a very short distance above a smooth ground, the air that gets between the wings and the ground below cannot flow on as easily as when the flying object is at a higher altitude.

The air now accumulates and thus exerts additional pressure on the wings, supplementing the lift that the wings already have.

Large seagulls, for example, like to take advantage of this welcome help to glide over the water for hours without much effort. The second result of this effect relates to the wake vortices that all aircraft drag behind them. In an aerofoil, there is always negative pressure on the upper side and positive pressure on the lower side. Therefore, the air not only tends to flow backwards on the wing, it also flows around the end of the wing to the upper side.

However, as the aircraft continues to speed in the meantime, it cannot do this continuously. Instead, it is swept away and now forms vortices that spread backwards from the edge of the wings, acting like a wake of the aircraft in the shape of an elongated cone. Wake vortices caused by large aircraft are strong enough to swirl smaller machines through the air. They also exert drag on the aircraft.

In the case of the ground effect, the phenomenon occurs that these vortices can only form to a much lesser extent in the immediate vicinity of the surface to the wing and are therefore much smaller than they are at altitude.

English fighter pilots discovered the ground effect

A photomontage showing a flying ship in the port of Hamburg. The FLYSHIP has been designed to adapt to the infrastructure of seaports. The bulky wings can be folded in. An auxiliary drive allows it to move slowly like a boat. Otherwise, the air jet from the propellers would quickly blow the St. Pauli Landing Bridges, which are busy with tourists, dry.

when they flew home from the French side of the Channel after an attack with only a little fuel left in their „Spitfires". Hardened pilots were able to fly very low over the Channel and thus greatly reduce fuel consumption. This was because the extra lift meant that the aircraft engine did not have to work as hard to maintain the same speed as at a higher altitude. Due to the ground effect, savings of up to 30 to 40 per cent can be achieved compared to a flight at a higher altitude. That is reason enough to take a closer look.

Conventional aircraft are not inherently set up to behave as stably in ground-effect flight as they do at altitude, because the point of attack of the lift near the ground on the wings shifts backwards.

Also, the nerve required to race along over the water at a speed of 250 km/h at a height of about one to two metres is quite considerable. So a technology had to be developed to make this kind of flying safer if one wanted to profit from the positive savings effect.

In Germany, after the war, the designer Dr. Alexander Lippisch was very interested in this subject and began extensive experimental work with the prototype „X-112", initially in the USA. His idea was to turn a triangular delta wing around and give it a slightly negative V-position. In addition, there was a T-tail rising high above the tail of the aircraft for altitude and lateral control, which absorbed the shift in the centre of gravity. The „X-112" flew stably right away and gave him the courage to build more.

In cooperation with the Rhein Flugzeugbau company, a whole series of experimental aircraft were built with funds from the German Ministry of Defence.

but not everything went smoothly. Accidents occurred that revealed the weaknesses of these devices. Like all seaplanes, they have to cope with waves on take-off and landing until they either fly or stand still. Therefore, they are at risk of rolling over or otherwise „breaking down".

It was not until the research work of Lippisch's successor, Hanno Fischer, who continued this work as the company's technical director, that the experimental carrier „Hoverwing 2VT" was able to develop a method to circumvent the take-off and landing emisery.

He combined the aircraft part of the ground effect

device with a catamaran hull that functioned like an SES (See there). The air cushion inside the catamaran tunnel, delimited on the outside by skirts, allowed the aircraft to take off and touch down with less drag and more protection against wave impact than its predecessors.

As Hanno Fischer always said, his aircraft floats

on the water before it actually took off. The air for the air cushion was branched off from the drive system of the „Hoverwing 2VT", so that an additional drive could be dispensed with. This new WIG flew extremely successfully.

Work is now underway to develop heavier flying boats for 40 to 80 passengers, which will cross the oceans at around 200 km/h. The fuel consumption per seat is to be

reduced. The fuel consumption per seat is to be considerably lower than that of a fast passenger ferry.

The highlight is that these flying ships will not be classified as aircraft, but as ships. This means that they do not have to meet such strict regulatory requirements as in aviation, and the vehicle can fit better into the traffic structure on the shipping lanes. If these devices are actually built, it will be tantamount to a revolution in the fast ferry industry. Here, too, the new electric drive could play a helping role.

The „Aquaglide 2 „ was an attempt by Russian developers to offer the „Ekranoplan" technology on the civilian market. It is too small for use at sea or on coasts.

Length:	10.70 m
Width:	5.90 m
Mass:	2,400 kg
Passengers:	4
Speed:	92 kn
Max. Wave height:	0.35 m

The „Ekranoplan"

No watercraft ever raced across the waves in a more spectacular way than the Soviet-Russian „Ekranoplan". The engineer Rostilav Aleekseev, who we have already come across in the Eastern hydrofoil development, developed this concept in the 1960s and offered it to the Soviet state. As a deserving and recognised scientist, he received very generous support from the military budget and began his work on ground-effect aircraft under the strictest secrecy.

Quite rightly, the military had recognised the potential of this type of vehicle as a means of transport for invading troops and as a weapons carrier. In contrast to Lippisch, Aleekseev pursued a more conventional path. He combined a slender fuselage with flat, low-lying rectangular wings and mounted the engines at the front and rear of the fuselage. The front ones were not only to support the thrust, but also to blow exhaust gases under the wings so that a cushion of air could form there even at low speed.

It was possible to make these huge aircraft, most of which far exceeded the dimensions of an Airbus A-380, fly. In the case of the largest Ekranoplan, the „KM", ten of the largest jet turbines available in Russia roared at full power to force the machine to take off. Once this was accomplished, it sped across the Caspian Sea at more than 200 knots (370 km/h). When, after a short flight, the coastline on the other side of the inland sea came into view, the test pilots had to turn around quickly because the sea was „running out" too fast.

The technical effort was as enormous as the energy input. Only a very complex autopilot was able to keep the aircraft under control. It did not have the natural stability caused by the inverted delta wings discovered by Dr. Lippisch. Another problem lay in the extremely high landing speed of about 200 km/h. The waves thereby set the flying boat in motion. The waves were so hard on the flying boats that dangerous material fatigue occurred. The concept stalled in development and Rostilav Aleekseev was withdrawn from development.

(Below) The „Lun", the second largest Ekranoplan ever flown, is bumming around somewhere on the Caspian Sea in 2010. At 73 metres long and with a take-off weight of about 380 tonnes, it is a giant compared to the patrol boats behind it, which are not exactly small either. Despite its speed of 550 km/h (296 kn), it had no future. Its use was simply too dangerous and too expensive.

(Right) The „Lun" was also flown armed according to its purpose. With its nuclear sea target missiles, it would theoretically have posed a great danger to NATO units.

Length:	73.1 m
Wing span:	43.9 m
Altitude:	19.8
Take-off mass:	400 to
Speed:	548 km/h

When looking at the dimensions, it is noticeable that this aircraft was slightly larger than the Airbus A-380. The „KM", on the other hand, was really the largest aircraft flown so far.

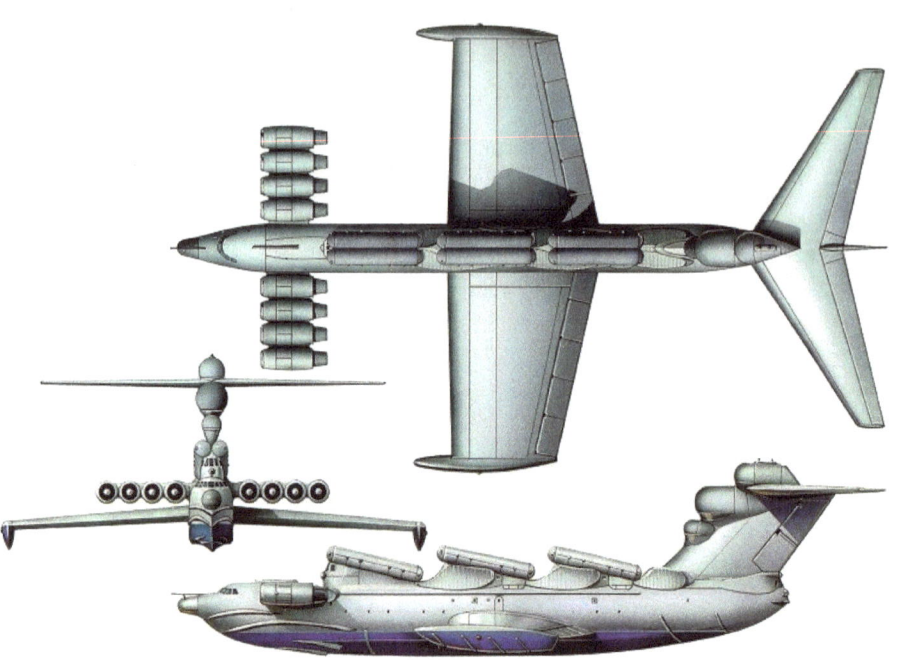

The US forces first noticed these experiments through unclear satellite photos. The admirals of the US Navy in particular were unpleasantly surprised, because the „Lun" could have easily transported cruise missiles in low-level flight undetected in the vicinity of American aircraft carrier battle groups. The reaction time of the fighter planes and defensive missiles that would have had to respond to this threat would have been insufficient. The Navy would have had to accept the heaviest losses. But this is pure theory. The Soviet state spent itself so much on these and other huge projects that it could never have built enough of these „monsters". After all, the people can't eat weapons.

When Aleekseev left and later died, the experiments continued, but they now concentrated on landing craft for transporting assault troops. Called the „Orlenok", it had a bow that folded to the side and could carry vehicles, troops and light tanks in its hull. A propeller turbine with over 14,000 hp was mounted on the tailplane for propulsion. In addition, the „Orlenok" received two jet turbines, which were placed in the bow. They were oriented downwards and slightly to the sides and blew their exhaust gases under the wings located in the middle. This allowed the vehicle to glide forward even slowly on land, but compared to a real hovercraft, this was just an awkward way to move.

The ground-effect aircraft, however, was successfully tested and would certainly have been mass-produced if the Soviet Union had continued to exist in its former glory after 1989. But as it was, there was soon no more money and the test vehicles have been gradually rotting away ever since.

There have been repeated attempts to market the „Orlenok" internationally as a means of civilian transport, but so far there has been no interest from potential operators. This form of ground effect vehicle is too risky and too expensive to operate. Since the 1990s, former members of the development team have been pursuing the project of constructing a small vehicle with four seats using the knowledge gained. But despite a successful flying prototype (see photo of the „Aquaglide 2" on p. 161), no prototype has yet been built.

on p. 161), no series production has yet begun.

Somewhere in Russia, a „monster" called „Spasatel" slumbers in a hangar, waiting for the day it is released. It is one of the largest aircraft ever built in the world. Since there is no more funding from the state and the economy, the builders have little more to do than regularly dust it off. The ground-effect form invented by Aleekseev has certainly achieved record-breaking and spectacular feats.

But since it could not solve the problem of stability and, moreover, the energy input was inflationary, this variety of ground effect technology will hardly be granted success.

All in all, the technology of ground-effect aircraft is now considered too risky for daily use. But this may change in the course of ever better electrical storage and drives, as the savings effect would then be more effective. In any case, this idea should not be dropped. Boeing is also developing its own giant WIG for cargo transport in the USA - financed by the military.

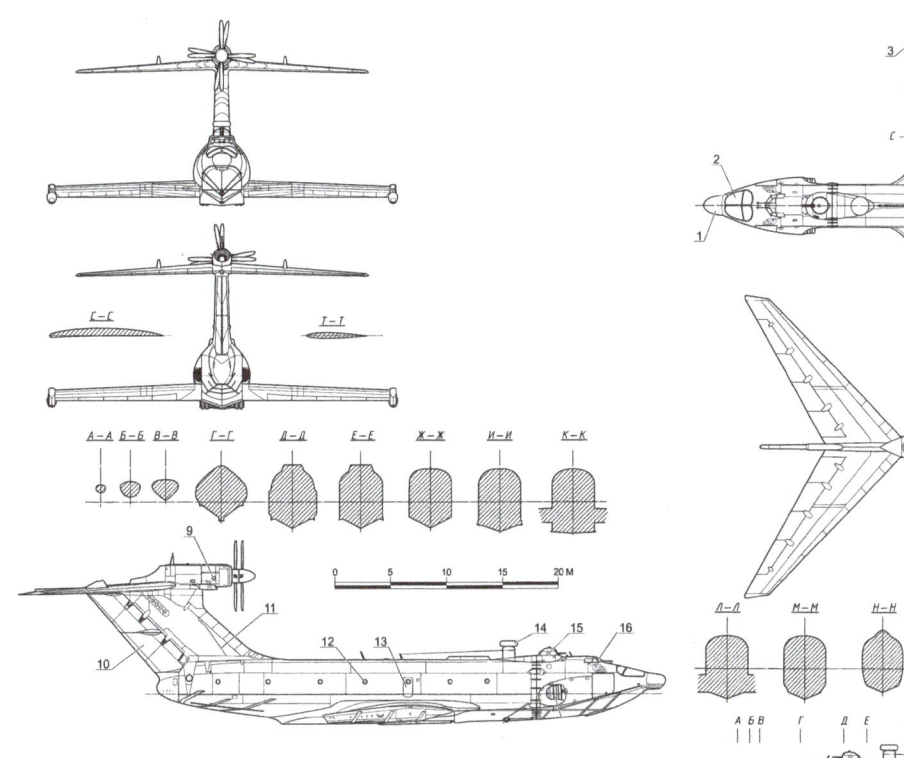

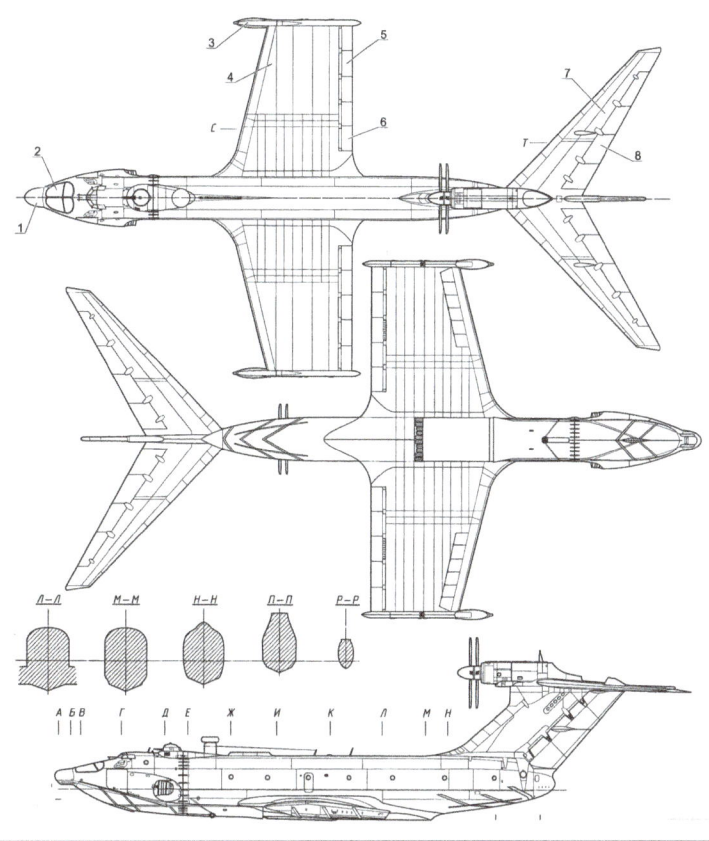

(Above and right) The „Orlyonok" is a later development that was significantly smaller than the „Lun". It flew successfully but was never built in series. It is exhibited in a Moscow museum.

Length:	58.5 m
Wing span:	31.5 m
Height:	16,5
Take-off mass:	140 to
Speed:	400 km/h
	(215 kn)
Propulsion:	1 x Kuznetsov NK12 (15,000 hp)
	2 x jet engine Kuznetsov NK8 (152 kN thrust)

12. High speed for model makers

(Above) Today it is a rare collecting piece: The small S.R.N. 6 model by Matchbox. The model was injection moulded from zinc.

For the modeller, high-speed ferries present a particular challenge. The complex hull shapes of catamarans, the wing construction of hydrofoils and the hovering function of hovercrafts cannot be reproduced on the model without technical calculations or the use of CAD programmes.

If a ship is scaled down, it has to be taken into account that the density of the water and the air as well as the power ratings for propellers, jet drives or fans cannot be scaled down in the same proportion. Therefore, buoyancy generating components - be it hulls, wings or lift fans must be completely recalculated if one does not want to experience a great disappointment with a remote control

model of a fast watercraft,.Fortunately, many other modellers have already taken these steps and some of them have published reference books that give precise information about the various problems and their solution. It is recommended by the author to order books of this kind mainly from English or American internet bookshops, as the corresponding offer in Germany is still a little sparsely populated.

On the other hand, many useful parts are already available in Germany, which can be helpful when building a model fast ferry. In the catalogues of the established suppliers in this country, complete jet engines are available that can impressively simulate the function and performance of the real engines.

Several kits of fast ships are also offered in specialised shops, they can be purchased as old kits or ready-built models in internet auction shops.

For the friend of polystyrene floor models, model kits of the „Finnjet" and other vessels are available from collectors or model builders. If, on the other hand, you want

(Left) A model of the police boat POT-3 already presented in this book. The model kit was offered by the now defunct company HEGI. It is still possible to buy unassembled kits from collectors on the internet.

to create your own fast ferry model, you should start with a monohull. Their hull design is similar to that of speed-boats and superyachts and is easier to design and build than a catamaran or hovercraft. However, with all models powered by water jets, it should be noted that the jets of the models must be dimensioned much larger in relation to the model than they are in the original.

This is because the power calculation of the model drive follows its own laws instead of being subject to the reduction factor. Decisive physical variables here are the friction of the water in the drive system and the weight of accumulators and electric motors. With hydrofoils, as with a model aeroplane, it is very important to weigh out the centre of gravity of the boat exactly, otherwise the model will either not take off or will always hit the water with its nose. Since there are now extremely light electric motors and batteries, it has become easy to equip models with too much power. This can cause a hydrofoil to lift too high out of the water and therefore virtually „crash". Hovercrafts are the most difficult model building projects because of the apron systems. Although the hulls of hovercrafts are

rather boxy and clumsily shaped and can also be easily influenced from hard foam blocks, the development of an apron that functions faithfully on the model sometimes drives builders to despair. The apron material must be very thin and yet resilient. The preferred materials are currently thin synthetic rubber or nylon films. The skirts must be cut and installed with extraordinary precision, as irregularities lead to air losses in the air cushion.

(Below/Above) For a long time there was a plastic stand model of the SR.N. 4 hovercraft from Airfix. The kit is of good quality and of-ten still available from web shops. The only thing it can't do is drive. The perfectly built model below was made by Richard J. Price. (UK)

13. Finally

The world of fast ferries is full of technical wonders. It is hardly possible to show all their diversity and the background of developments in just over 160 pages. The author hopes that readers have been provided with an informative overview. Those who would like more in-depth information should note the following names and addresses. It should be noted that this small list is a selection according to the author's preference and does not claim to be complete. In addition, addresses and web addresses change from time to time. All important media appear in English, as fast ferry shipping is hardly known in Germany.

Websites (examples):

Ship Technology, UK, (www.ship-technology.com)

Industry overview of newbuildings

Ship Publications, NO, (www.shipping-publ.no)

Link list to important publications

International Hydrofoil Society (www.foils.org)

The international society of all hydrofoil fans and technologists

Classic Fast Ferries, DK, (www.classicfastferries.com)
High quality internet fanzine by a Danish journalist, unfortunately not updated at the moment, but there is an issue archive with PDF files for download. Certainly one of the best sources for fans in English.

Imprint

Print on demand
July 2023
Copyright © 2020/2023 by Windfang-Medien, Christof Schramm
28357 Bremen (Germany) ISBN 9798851274497

www.ingramcontent.com/pod-product-compliance
Lightning Source LLC
Chambersburg PA
CBHW041457280526
45792CB00004B/1040